AGRICULTURE ISSUES AND POLICIES

FEED MARKET DYNAMICS AND U.S. LIVESTOCK IMPLICATIONS

AGRICULTURE ISSUES AND POLICIES

Additional books in this series can be found on Nova's website under the Series tab.

Additional E-books in this series can be found on Nova's website under the E-book tab.

CONTENTS

PREFACE

This book provides a review of the current feed market dynamics including the major emerging issues and their implications for the U.S. livestock sector and Congress and background information on the market structure of the U.S. feed grain sector.

Chapter 1- The U.S. livestock and poultry sector is presently confronting near-record high feed costs driven by the convergence of both current tight supply prospects and long-run market forces. The U.S. livestock sector, where feed costs account for 50% to 80% of cash operating expenses, has seen its profit margins steadily squeezed and its share of U.S. agricultural cash receipts decline. Since June 2010 feed cost increases have outpaced livestock price increases, squeezing the profitability of livestock and poultry producers. When profit margins turn unfavorable, producers are more likely to retain fewer breeding animals or to liquidate herds. Near-term liquidation means smaller future herds or flocks and ultimately higher retail prices for beef, pork, poultry meat, eggs, and dairy products. Persistent upward feed price movements could result in substantial and long- lasting consequences for livestock product prices throughout the marketing chain.

Chapter 2- The byproducts of making ethanol, sweeteners, syrups, and oils used to be considered less valuable than the primary products. But the increased livestock-feed market for such byproducts in the past few years has switched that perception to one of the ethanol industry making grain-based "co-products" that have market value separate from the primary products. Co-products such as dried distiller's grains, corn gluten feed, corn gluten meal, corn oil, solubles, and brewer's grains have become economically viable components, along with traditional ingredients (such as corn, soybean meal, and urea), in feed rations. The co-products have limitations, such as variable moisture content, product availability, nutrient excesses or deficiencies, and nutrient variability. These limitations affect how they must be handled and stored and how much they cost feed buyers. Dried distiller's grains are more amenable to pelleting and other bulk-handling methods than other co-products, which gives them an advantage in international markets.

Chapter 3- Animal and poultry disease outbreaks often lead to new or amended policies and regulations. The economic effects induced by these policies can be much greater and much longer lasting than the immediate effect of the disease outbreak alone. Using Bovine Spongiform Encephalopathy (B SE) as an example, this paper demonstrates the pervasiveness of the effects of restrictive feed policies and regulations, particularly as they relate to meat and bone meal and other protein feeds. Costs evaluated include those assumed by consumers via changes in supplies of secondary and final products; environmental costs associated with disposal of hazardous materials; lost value of products to the rendering industry, including a decline in value of meat and bone meal; and costs of supply disruptions and substitutions within the feed market sector. Benefits from new or amended policies accrue but are not easily measured.

Chapter 4- Livestock producers in 2008 have seen sharply higher feed costs, fueled by competing use demands for corn and soybeans and by higher energy prices. Some analysts argue that current public policies, including financial incentives that divert corn from feed uses into ethanol production, exacerbated if not caused these higher costs. Other factors, which some believe to be at least as significant, include crop production declines due to weather, and higher global demand for commodities. Proposed options aimed at easing the impacts of higher feed costs include changes in ethanol incentives, use of conservation land for forage use, and direct aid to producers.

In: Feed Market Dynamics …
Editors A. Stewart and K. Gray

ISBN: 978-1-61942-272-8

Chapter 1

U.S. LIVESTOCK AND POULTRY FEED USE AND AVAILABILITY: BACKGROUND AND EMERGING ISSUES

Randy Schnepf

SUMMARY

The U.S. livestock and poultry sector is presently confronting near-record high feed costs driven by the convergence of both current tight supply prospects and long-run market forces. The U.S. livestock sector, where feed costs account for 50% to 80% of cash operating expenses, has seen its profit margins steadily squeezed and its share of U.S. agricultural cash receipts decline. Since June 2010 feed cost increases have outpaced livestock price increases, squeezing the profitability of livestock and poultry producers. When profit margins turn unfavorable, producers are more likely to retain fewer breeding animals or to liquidate herds. Near-term liquidation means smaller future herds or flocks and ultimately higher retail prices for beef, pork, poultry meat, eggs, and dairy products. Persistent upward feed price movements could result in substantial and long- lasting consequences for livestock product prices throughout the marketing chain.

USDA forecasts that U.S. corn stocks will approach historic low levels relative to demand by the end of summer 2011. In addition, the current crop outlook remains uncertain due to late planting in the Northern Plains and Eastern Corn Belt and a severe drought that has dried up pasture lands from the Southern Plains into the Southeast, exposing cattle producers to expensive commercial feed markets.

From a longer-term perspective, U.S. feed grain demand has exceeded production in all but one year since 2004. This persistent trend is primarily the result of five factors: rapid growth of U.S. corn-based ethanol production (whose share of the U.S. corn crop in 2011 is forecast to exceed feed use for the first time), limited supply of available U.S. cropland to expand production, prolonged weakness of the U.S. dollar which has made U.S. agricultural exports competitive in foreign markets despite high prices, strong income growth in China and other international markets which has increased demand for livestock products and the feedstuffs (feed grains and protein meals) needed to produce them, and a substantial decline in the price responsiveness of both supply and demand in agricultural commodity markets. The convergence of these factors has resulted in falling grain and oilseed stocks, record or near-record prices for most feedstuffs— grains, oilseeds, hay, and pasture—in 2011, and increasingly volatile commodity prices. These factors are expected to persist for several more years, thus maintaining strong demand for feed grains and strong upward pressure on prices for all commodities that compete for farm land.

Issues surrounding feed use and availability are likely to be of growing interest to Congress as the House and Senate Agriculture Committees monitor the financial health and well-being of the U.S. livestock and poultry sectors, as well as any future ramifications for retail food price inflation. In addition, Congress, along with energy and agricultural market participants, will closely follow any further acceleration of unanticipated side effects on the U.S. livestock sector from continued corn use for ethanol production. Finally, feed availability and associated market price volatility are likely to play a significant role in the next farm bill debate, as the current omnibus farm bill expires in 2012.

Feed market dynamics and what they mean for the U.S. livestock sector depend on many factors. This report provides a review of the current feed market dynamic including the major emerging issues mentioned above and their implications for the U.S. livestock sector and Congress. In addition, background information on the market structure of the U.S. feed grain sector is contained in an appendix.

INTRODUCTION

The U.S. livestock and poultry sector is presently confronting near-record-high feed costs. These high feed costs—which account for 50% to 80% of cash operating expenses of livestock and poultry producers—pose significant consequences for the sector. First, feed cost increases have steadily squeezed the profit margins of livestock and poultry producers, putting the sector under increasing financial pressure. Second, strong grain prices have decreased the value-share of the livestock and poultry sector relative to the crop sector, as the livestock share of U.S. agricultural cash receipts has declined sharply from 51% during the 2000-2005 period to 43% since 2008 (Figure 1). Finally, when profit margins turn unfavorable, producers are more likely to retain fewer breeding animals or to liquidate herds.

Near-term liquidation means smaller future herds or flocks and ultimately higher retail prices for beef, pork, poultry meat, eggs, and dairy products. Persistent upward feed price movements could result in substantial and long-lasting consequences for livestock product prices throughout the marketing chain.

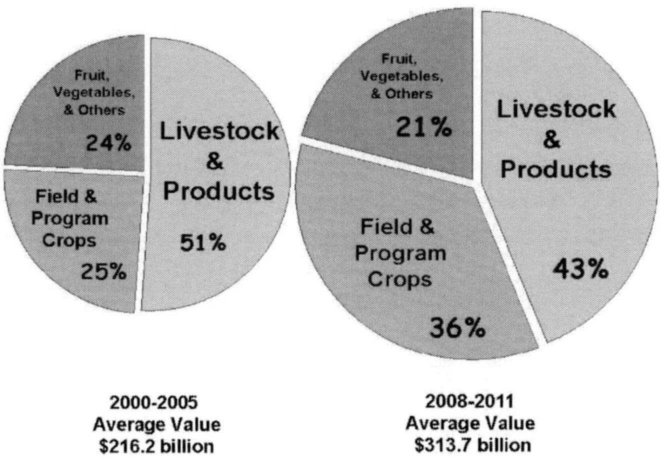

Source: Economic Research Service (ERS), USDA, February 14, 2011.

Note: Calculations are based on total annual cash receipts of U.S. agricultural production activities.

Figure 1. U.S. Agriculture Sector Value Shares.

This report describes the forces that are behind declining domestic feed availability and rising feed costs, and what they mean for the U.S. livestock sector and Congress. It also provides background information on the structure of the U.S. feed grain sector in an Appendix.

FEED MARKET DYNAMICS

The major commercial livestock species in the United States include beef cattle, dairy cows, hogs, and poultry. Three principal types of feedstuffs are used in livestock and poultry rations— energy feeds (primarily from feed grains such as corn and sorghum), protein feeds (primarily from oilseed meals), and roughage (primarily from grazing pasture and rangeland or from cut hay and silage).[1]

Finding the Right Feed Nutrient Balance

Livestock and poultry producers are always looking for the most cost-effective feed sources to achieve maximum weight gain, egg production, and milk output from their herds and flocks. For overall animal health and productivity all commercial livestock and poultry species require a balanced nutrient intake of energy, protein, and roughage, supplemented by essential vitamins, minerals, and other minor additives.[2] Fortunately, producers can use a wide variety of natural feed ingredients to satisfy the nutritional requirement of their livestock and poultry.

The choice of feedstuff depends critically on the type of livestock or poultry production activity, since each animal or bird species has different abilities to use any or all of the various feed types based on their digestive capabilities (e.g., ruminant versus monogastric versus avian), as well as the age, sex, breed, health, and reproductive status of the individual animals in a herd or flock. As a result, livestock producers must carefully balance their animals' nutritional needs with the cost and availability of feed supplies.

In particular, ruminant animals such as cattle, sheep, and goats are able to break down and digest the cellulose in high-fiber roughages. In contrast, monogastric species (e.g., hogs) or avian species (e.g., poultry) are not able to easily digest cellulosic material and will become ill if their feed rations are too high in fiber. As a result, swine and poultry producers rarely use pastures and ranges in their feeding operations. Instead, swine and poultry are the major consumers of both feed grains and protein meals, while cattle and dairy cows are the major consumers of roughage. For example, the hog industry is such a voracious consumer of the U.S. corn crop that a mapping of U.S. hog inventories very closely mirrors a mapping of corn production (Figure A-1). Similarly, the rapid growth of U.S. poultry meat production which has occurred in recent decades is one of the primary factors behind strong growth in U.S. protein meal consumption.

Feed Costs Dominate Livestock and Poultry Production

Livestock and poultry producers have some flexibility in which feeds they use based on the specific nutritional needs of their livestock or birds, and the relative availability and price of the various feed alternatives. Supply and demand changes within and across the three major feedstuff categories impact each major species differently. In general, feed costs account for 50% to 80% of cash operating expenses for most livestock and poultry operations.

Table 1. U.S. Operating Cost Shares for Livestock and Poultry Production Activities, 2009-2010 Averages

Activity	Feed Costs	Other Costs	Total Operating Costs
Milk	76.9%	23.1%	100%
Cow-Calf	69.5%	30.5%	100%
Hogs: All	55.4%	44.6%	100%
Hogs: Farrow to Finish	84.1%	15.9%	100%
Hogs: Feeder to Finish	52.5%	47.5%	100%
Poultry	70.0% [a]	30.0%	100%

Source: "Commodity Costs and Returns Data," *Data Sets,* ERS, USDA, updated July 6, 2011.

Notes: Farrow to Finish is production from birth until marketed for slaughter at about 225 pounds. Feeder to Finish is production from about 50 pounds until marketed for slaughter. A farrow is a litter of pigs at birth.

a. Representative poultry

b. cost-of-production data are not available; however, the cited figures are shares of live cost, given by Dan King, Virginia Poultry Federation, as part of his testimony at a House Committee on Agriculture, Hearing on the "State of the Poultry Industry" for the Subcommittee on Livestock, Dairy, and Poultry, April 13, 2011.

Changes in feed grain supply, primarily due to weather variations, cause most of the short-term market price changes. The total supply of feeds also depends on many other factors including producer planting choices, agricultural policies, climate, soil type, availability of processing and storage facilities, and international market conditions. Comparative market prices derived from the demand for feed by the various livestock species are also of paramount importance.

Producer Responses to Feed Market Conditions

Because animal and bird nutrient requirements can be met from a variety of feed sources, livestock and poultry feeders are sensitive to comparative price changes among feedstuffs (Figure 2). They are also very sensitive to relative changes between feed prices and the prices of their livestock products (Figure 3). Often, the proximity of livestock and poultry operations to a particular feed source is as important in the feed choice as are relative feed prices and the specific nutrient requirements of each animal or poultry species. Because nutrient needs and feed availabilities usually shift with the seasons, a producer is constantly challenged to plan three to six months or more in advance to avoid being caught in a cost-price squeeze (i.e., product prices fall while feed costs rise) by an unforgiving marketplace.

Prices for various feed grain are linked by their relative feed values and their competition for cropland. As a result, feed grain prices are highly correlated and tend to respond to the same market factors such as weather conditions and changes in international supply and demand. However, because of corn's dominant role in U.S. and world feed grain markets, the price of other feedstuffs tends to follow corn price changes (Figure 2).[3]

Swine and poultry producers are generally more corn-dependent than cattle producers, at least until the cattle have been placed on feed as the final stage prior to marketing. When the hog-corn or steer-corn price ratio turns unfavorable—either due to a corn price upswing, a meat price downturn or some combination—producers are more likely to market their stock earlier to minimize feed costs or to reduce the number of breeding animals that they retain to minimize future herd sizes and feed needs. Depending on the stage of the livestock cycle, changes in breeding activities can affect future market conditions. It can take several years for cattle producers to fully respond to grain market conditions due to the lengthy gestation (nine months) and growth cycle of cattle. Hog producers can respond within five to seven months due to shorter gestation and growth cycles; while broiler and egg producers can modify their operations within a few weeks.

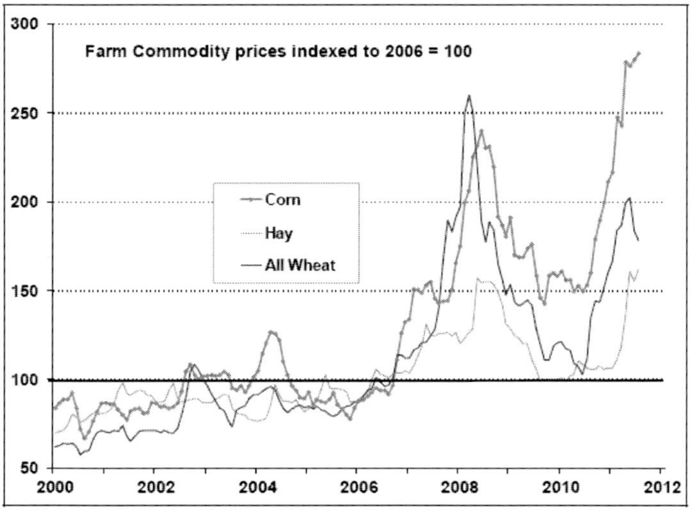

Source: Calculations by CRS using data from NASS, USDA, July 31, 2011.

Figure 2. U.S. Monthly Farm Prices Received for Corn, Wheat, and Hay Since 2000.

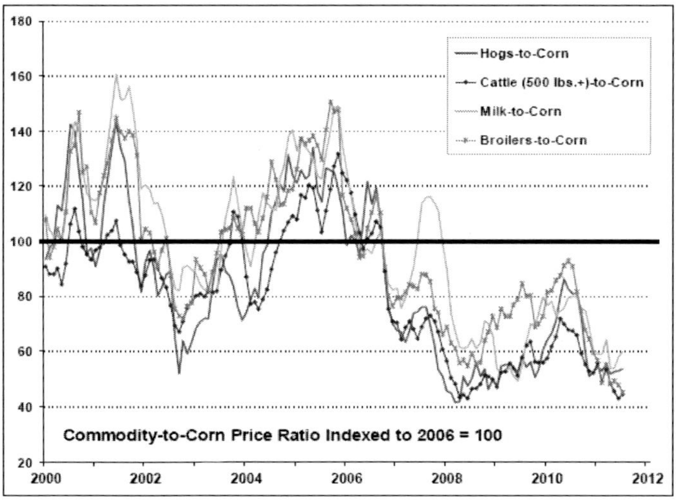

Source: Calculations by CRS using data from NASS, USDA, July 31, 2011.

Figure 3. Livestock-to-Corn Price Ratios Have Trended Downward Since 2006.

CURRENT AND EMERGING ISSUES FOR LIVESTOCK FEEDSTUFFS

High feed costs are being driven by the convergence of both current tight supply prospects and long-run market forces. With respect to current conditions, USDA forecasts that U.S. corn stocks will approach historic low levels relative to demand by the end of summer 2011. In addition, uncertainty about the current crop outlook, due to late planting in the Northern Plains and Eastern Corn Belt and a severe drought that has spread from the Southern Plains into the Southeast, has dried up pasture lands, exposing cattle producers to expensive commercial feed markets.

From a longer-term perspective, U.S. feed grain demand has exceeded production in all but one year since 2004. This persistent trend is primarily the result of five factors:

- rapid growth of U.S. corn-based ethanol production (whose share of the U.S. corn crop in 2011 is forecast to exceed feed use for the first time),
- limited supply of available U.S. cropland to expand production,
- prolonged weakness of the U.S. dollar which has made U.S. agricultural exports very competitive despite high commodity prices,
- strong income growth in China and other international markets which has increased demand for livestock products and the feed grains and protein meal needed to produce them, and
- a decline in the responsiveness of feed demand to higher prices—the result of tight stocks, limited cropland, the increased use of corn for ethanol production, and the increasing industrialization of hog and poultry production.[4]

These factors are expected to persist for several more years, thus maintaining strong demand for feed grains and strong upward pressure on prices for all commodities that compete for farm land. The result of these convergent forces has been falling grain and oilseed stocks, record or near-record prices for most feedstuffs—grains, oilseeds, hay, and pasture—in 2011, and increasingly volatile feed grain prices due to tight supplies and the declining demand responsiveness.

Current Corn Stocks-to-Use Levels Near Historic Low

Market observers typically monitor the relative market tightness of a commodity with the ratio of end-of-season stocks to the total use during the season. Season-ending stocks of an annually produced commodity such as corn summarize the effects of both supply and demand factors during the marketing year.[5] A certain amount of stocks at the end of the marketing year are necessary to provide a continuous flow of grain to processors and exporters before the new crop is harvested. These stocks are referred to as pipeline supplies. Although there is no hard and fast rule on what volume of stocks represents minimum pipeline levels for the major grain and oilseed crops, a stocks-to-use ratio below 7% clearly raises market concerns. Whenever USDA ending stock projections approach this level, market prices become very sensitive to unexpected market news and prices tend to be more volatile than during periods of abundant stocks.

According to USDA, by September 1, 2011 (end of the 2010/2011 corn marketing year), the ratio of U.S. corn stocks to use is expected to be 7.1%, the lowest level since 1996 when the ratio fell to 5% (Figure 4).[6] USDA forecasts that the ending-stocks-to-use ratio will decline further to 5.4% in 2012. The decline in U.S. corn ending stocks was precipitated by a surge in demand (up 23%) relative to production (up 9%) since 2004.[7] The result has been a sharp decline in the level of domestic corn stocks relative to use.

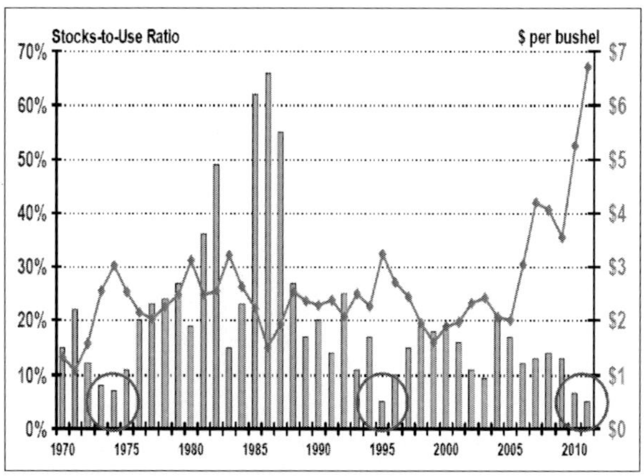

Source: NASS and WAOB, USDA, August 11, 2011.
Notes: 2011/2012 data points are USDA forecasts.

Figure 4. U.S. Corn Season Average Farm Price versus Ending Stocks-to-Use Ratio.

Severe Drought Limits Feed Alternatives for Affected Regions

Drought conditions that emerged over the winter in the Southern Plains have worsened considerably into the summer months, and have extended eastward across the entire Southeast and westward into New Mexico and Arizona (Figure A-2). The severe drought has reduced pasture and rangeland grazing opportunities in the affected regions, making livestock producers more dependent on commercial feed markets. In addition, it will likely reduce this year's grain crops in the affected regions, further increasing the livestock sector's dependence on feedstuffs from outside the region and raising production costs.

The period from October 2010 through June 2011 was the driest on record in Texas.[8] National Weather Service officials predict that the severe drought conditions in Texas and Oklahoma are likely to persist into early next year, and it will likely take months and a wet winter for the pasture land to recover according to a cattle market professional.[9] This would be crucial because if the drought persists into October, it will likely result in lower winter wheat plantings in Texas and Oklahoma.[10] A decline in winter wheat plantings would be a further blow to ranchers, who would lose key pasture lands. Southern Plains cattle often are permitted to graze on early growth of winter wheat through the winter and into early spring. At that time, depending on the relationship between livestock and wheat prices, either the cattle are allowed to continue grazing and no wheat crop is produced, or they are removed and the wheat is allowed to fill out for eventual harvest.

Ethanol's Growing Share of U.S. Corn Usage

The U.S. agricultural landscape has changed dramatically since 2004 with the ascendancy of the U.S. ethanol sector. Most U.S. ethanol is produced from corn starch making it a major source of demand for corn. As U.S. ethanol production has grown in recent years, so too has its usage share of the annual corn crop (Figure 5). In 2004/2005, national ethanol production used 1.3 billion bushels and accounted for 12% of total U.S. corn use; in 2010/2011 it used 5.1 billion bushels (or about 38% of total use) to produce an estimated 13.6 billion gallons of ethanol.[11] Corn ethanol advocates argue that nearly 30% of corn used for ethanol is returned to livestock feed markets as high-valued distillers grains and solubles (DGS).[12] When corn use estimates are adjusted for DGS, the 2010/2011 corn-for-ethanol use share is 29% compared with a 9% share in 2004/2005 (Figure 6).

Ethanol Corn Use Has Cut into Feed Supplies

Once an ethanol plant has been built and its fixed capital investments have been made, the plant has to cover only its variable operating expenses to remain operating in the short term. Strong international crude oil prices help to ensure continued profitability in the ethanol sector despite historically high corn prices. As a result, the ethanol sector is the likely buyer of first choice in acquiring local corn at the expense of other users, including both the livestock and export sectors (Figure 6).

Even when DGS supplies are included with corn and other feed grains, total feed grains consumed by livestock and poultry have fallen 11% during the past seven years, from 7.1 billion bushels in 2004/2005 to a forecasted 6.3 billion bushels in 2011/2012 (Figure 7).[13] Thus, feed markets have tightened considerably and total feed use has declined since the rapid expansion of the U.S. ethanol industry began in 2005.[14]

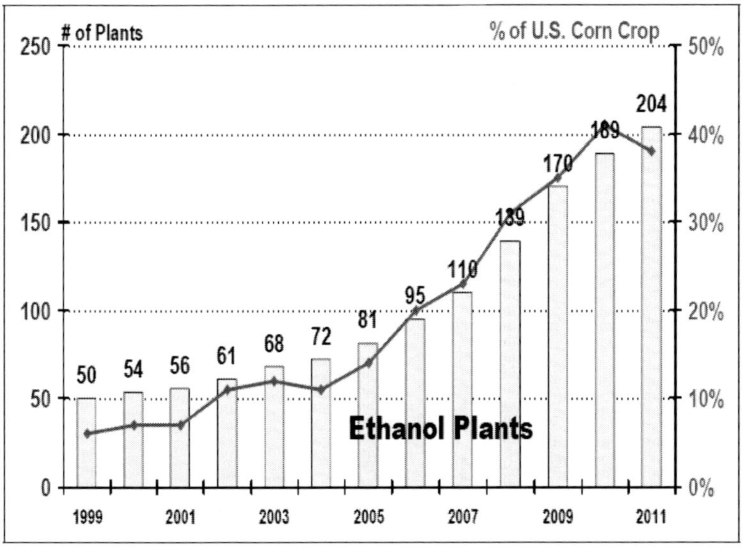

Source: Renewable Fuels Associations and WAOB, USDA (corn usage data for 2011 is a forecast).

Figure 5. Count of U.S. Ethanol Plants and Ethanol's Share of U.S. Corn Crop.

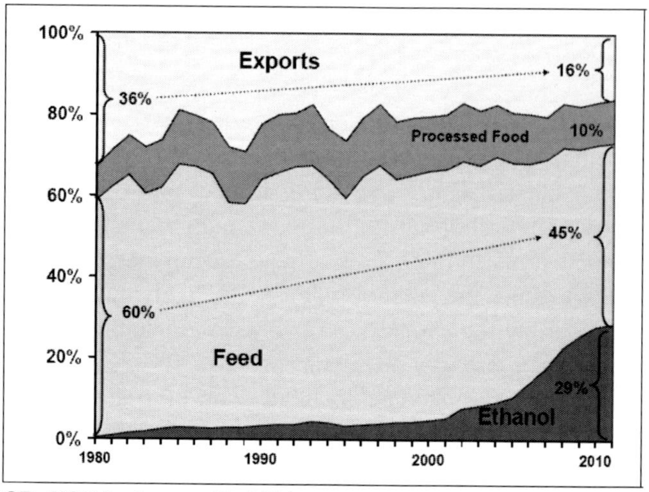

Source: WAOB, USDA, August 11, 2011.

Note: Corn-for-ethanol, corn-for-feed, and corn-for-export usage totals are adjusted to include the distillers grains and solubles (DGS) in corn-equivalent units used for each. Pre-adjustment shares for 2011 are 38%, 38%, and 14%, respectively. Note that Figure 6 compares total-use shares; whereas Figure 5 expresses the usage rates as shares of annual crop production.

Figure 6. U.S. Corn Total-Use Shares (%) by Activity, Since 1 980, Adjusted for DGS.

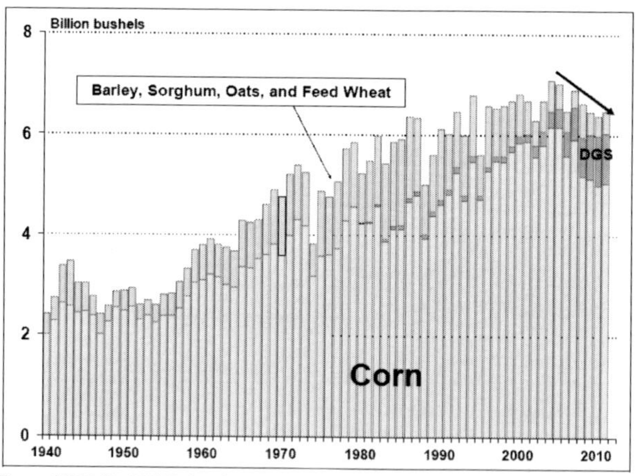

Source: WAOB and ERS, USDA, August 11, 2011.

Notes: Distillers grains and solubles (DGS) are in corn-bushel equivalents.

Figure 7. Total Feed Grains and DGS Fed to Livestock and Poultry.

U.S. Biofuels Policy Has Created Link Between Prices of Crude Oil and Corn

A key result from the dramatic expansion of U.S. ethanol production has been the development of a strong positive link between the prices of crude oil and corn.[15] In fuel markets, ethanol is a complement to gasoline up to federally mandated levels under the Renewable Fuel Standard (RFS),[16] and a competitor with gasoline above mandated levels. Increases in crude oil prices lead to increases in gasoline prices which, in turn, make ethanol more profitable to produce thus increasing the demand for corn.

Ethanol production from corn has nearly reached the maximum mandated level of 15 billion gallons by 2015 under the RFS, after which corn for ethanol demand is expected to slow dramatically.[17] Over time, as U.S. corn yields continue to expand, growth in corn production is expected to match or exceed growth in ethanol demand for corn—the eventual balance between ethanol production and corn demand will depend on many unknown market factors including the price of crude oil and the U.S. domestic blend wall.

In addition to the support from the RFS, the outlook for corn-based ethanol production remains strong due to two additional factors. First, the slow start-up of the cellulosic ethanol industry serves to keep both the biofuels and renewable-energy focus on corn-based ethanol because it is essentially "the only game in town." Second, current congressional interest in finding budget savings has led to several proposals to trim back or eliminate federal support for the development of new biofuels, including cellulosic ethanol, further reinforcing the emphasis on existing corn-based ethanol technology.

International Demand Growth Led by China

Since 2004, global demand for wheat and coarse grains has risen by over 15% while production has grown by only 10%. China, with over one-fifth of the world's population, is the primary driver behind the surge in international demand for livestock feedstuffs, primarily in the form of soybeans imported for both protein meal and oil.

Consumer demand for meat and other livestock products is highly correlated with income growth. China has had nearly 10% annual GDP growth for the past 30 years.[18] This tremendous economic growth has pulled millions of people out of poverty and into higher income strata. According to research by a market analysis firm, about 20% of China's population is upper-middle income or above (approximately 260 million people); that share is expected to increase to 30% by 2015 and almost 70% by 2025.[19]

Similar to national GDP growth, China's feed industry has also been growing at nearly 10% annually. However, limited domestic availability of both arable land and water has compelled China to turn increasingly to the international marketplace to meet its domestic needs and fend off domestic food-price inflation. For example, China's share of world soybean imports has risen dramatically from negligible imports in the early 1990s to nearly a 60% share of world soybean imports by 2009 (Figure 8). Stock building accounts for a substantial share of China's soybean imports.[20] USDA estimates that China's sobyean stocks now account for a relatively high 23% of usage. As a result, it is expected that future import and demand increases will slow down from the recent three-year expansion rate of 12% and follow the slower rate of income growth in future years (projected at 8.6% in 2011 declining to about 6% by 2025).

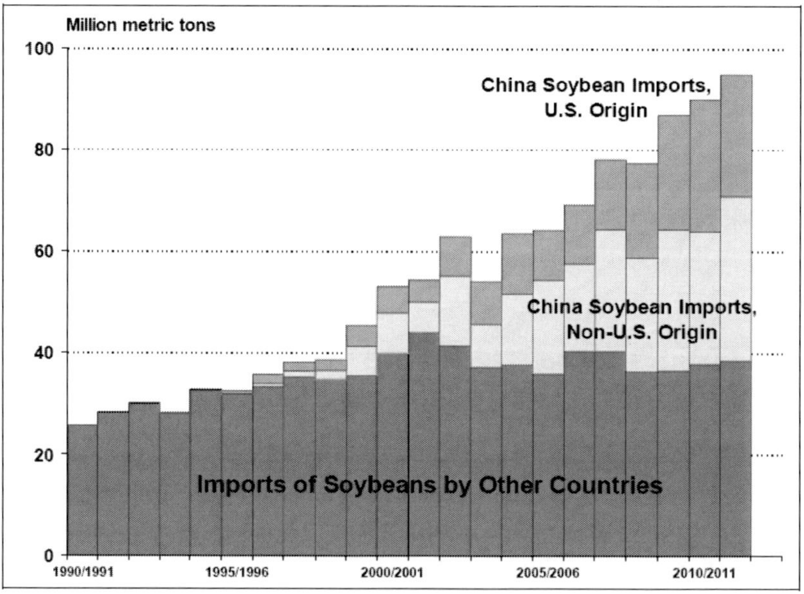

Source: PSD database, FAS, USDA, July 12, 2011.

Figure 8. World Soybean Imports Since 1990/1991.

Recent purchases of U.S. corn, along with market reports of the potential for large future corn imports, have heightened concerns that China may replicate its heavy reliance on soybean imports in the world corn market. China became a net corn importer in 2010, after 15 years of net exports, when it purchased 1.6 million metric tons (mmt) from the international market. During the first half of 2011, China had already bought 3.7 mmt of corn from the United States, and some are predicting that China's full year purchases will reach 5 mmt.[21]

USDA's recent long-run baseline study projects that strong income growth coupled with young, expanding populations in most major foreign markets—including Brazil, China, India, and Indonesia, as well as the rest of Asia, the Middle East, and South America—will continue through the 10-year outlook period.[22] The resulting increases in per capita income are expected to continue to generate strong demand for livestock and poultry products and the feed grains and protein meals needed to produce them throughout the period. As a result, grain and oilseed prices are projected to remain strong and producers are likely to be challenged to keep pace with demand.

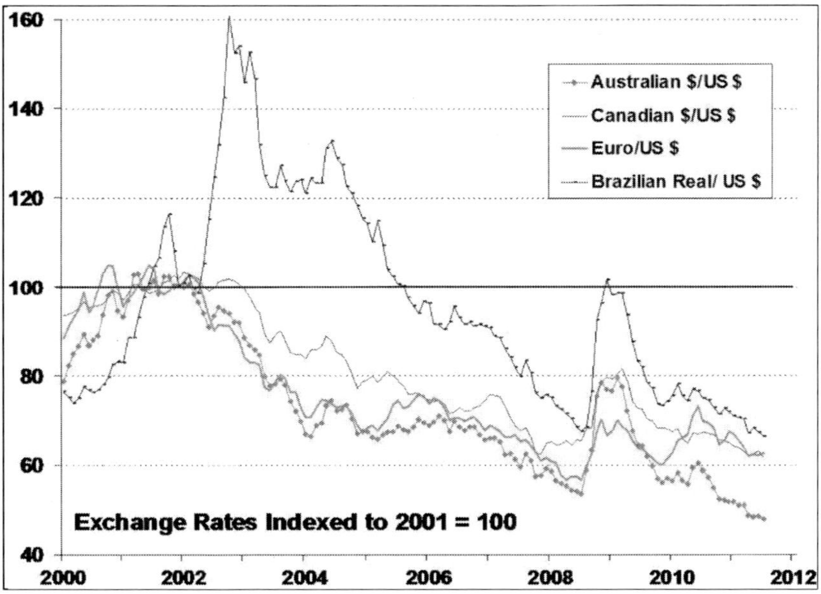

Source: Pacifica Exchange Rate Service, at http://fx.sauder.ubc.ca/data.html.

Figure 9. U.S. Dollar Versus Major Export Competitor Currencies.

Weak U.S. Dollar Makes Exports Attractive

U.S. feed grain exports have been facilitated by persistent weakness of the U.S. dollar against important foreign currencies—since 2001 the U.S. dollar has fallen substantially against major grain export competitors (Figure 9) Australia (-52%), Brazil (-34%), Canada (-38%), and the European Union (-37%), and importers (Figure 10) Japan (-35%) and South Korea (-18%). This strong and persistent decline in the U.S. dollar has made U.S. agricultural products more price competitive in international markets and helped to offset the strong run-up in commodity prices that has occurred in recent years. As a result, international demand for U.S. feed grains has remained robust despite high commodity prices, thus exacerbating already high prices.

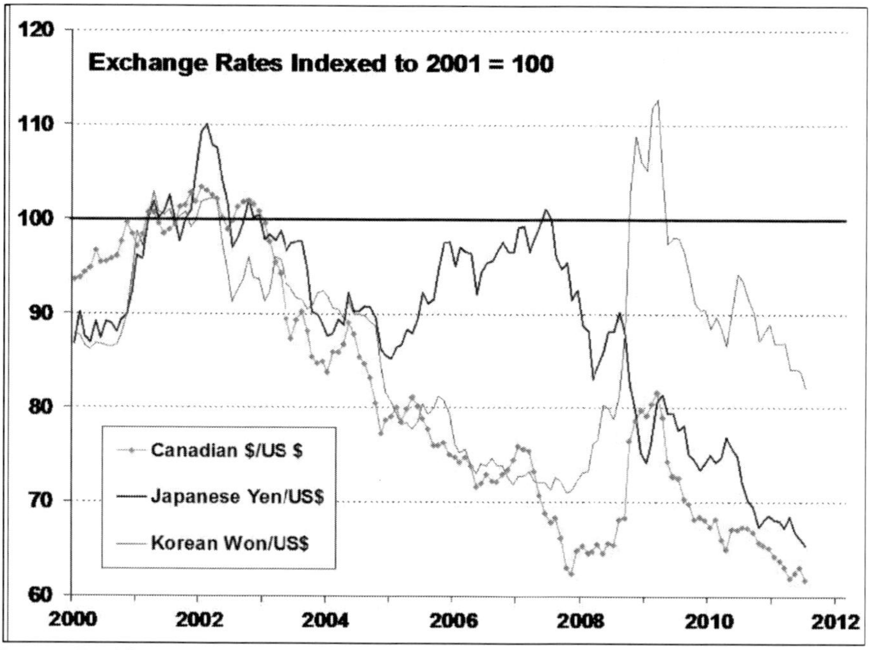

Source: Pacifica Exchange Rate Service, at http://fx.sauder.ubc.ca/data.html.

Figure 10. U.S. Dollar Versus Major Agricultural Importer Currencies.

Emerging U.S. Crop Land Constraint Puts Focus on Yield Increases

Growth in U.S. production of feed grains, oilseeds, and hay traditionally depends on yield and area developments. However, land and water constraints are beginning to place effective limits on the expansion of pasture and cropland in the United States. Total U.S. planted area to the major field crops (corn, soybeans, wheat, other feed grains, minor oilseeds, and upland cotton) appears to have peaked in 1981 at about 302 million acres (Figure 11). Since the early 1980s, substantial tracts of cropland have been idled temporarily under federal set-aside programs or removed from production under long-term Conservation Reserve Program (CRP) contracts.

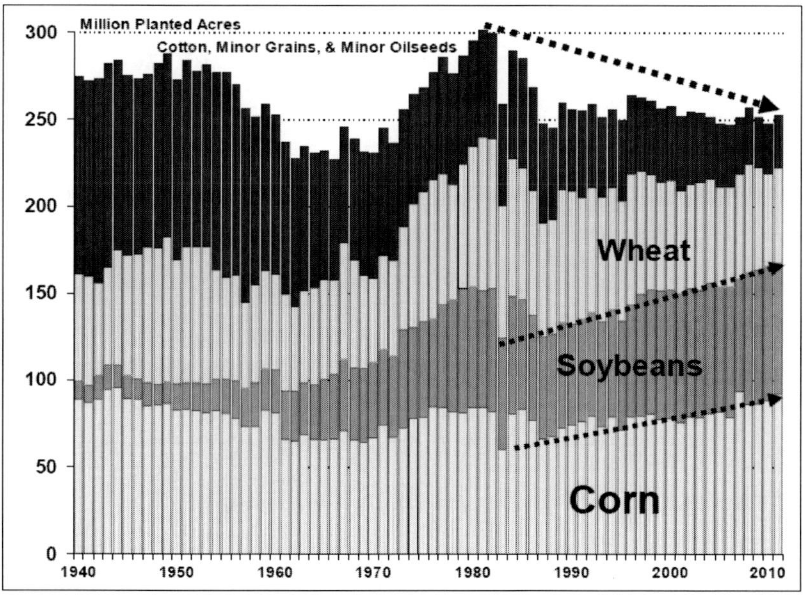

Source: NASS, USDA, August 11, 2011.

Note: Data exclude acreage idled under federal set-aside programs or the CRP. Data represent summation across different marketing years—Sept.-Aug. for corn, sorghum, and soybeans; June-May for wheat, barley, oats, and rye; Aug.-July for cotton.

Figure 11. Feed Grain, Oilseeds, and Cotton Planted Area Since 1940.

Since 1990, total planted acres for major field crops has hovered in the range of about 250 million acres; however, corn and soybeans have clearly gained acres at the expense of the other crops. Set-aside programs ended with the 1996 farm bill (P.L. 104-127; however, as of March 2011 over 31 million acres of farmland were enrolled in the CRP. There has been an active debate between CRP defenders who want to preserve its size, and those groups either seeking access to more cropland or concerned about the cost of supporting such a large land reserve.[23] Barring termination of CRP, future crop production growth will likely depend increasingly on yield gains which, in turn, hinge crucially on both weather patterns (which are increasingly fraught with uncertainty related to global climate change) and the level and success of new investments in agricultural research.

Declining Feed Demand Responsiveness to High Prices

In economic theory, a rise in the price of a good results in a decrease in the quantity demand of that good. The degree to which the demand falls is described as its "responsiveness." Theory suggests that a lack of viable substitutes is often an important reason for low demand price responsiveness. Also, the industrialization of agricultural processes (e.g., a corn-based biofuels plant or a large-scale hog-feeding operation) decreases the responsiveness to a price change in a single factor of production (e.g., corn).

A recent study suggests that feed grain demand is becoming less responsive to agricultural commodity price increases.[24] This decline in price responsiveness is itself an important factor in causing prices to be both higher and more volatile in response to demand shocks. The study cites five sources for the decreased price responsiveness of feed grain demand:

- biofuels policy which reinforces ethanol demand for corn over other uses,
- high livestock prices since 2009 which have contributed to strong feed demand despite high feed prices,
- very tight feed grain stocks relative to use,
- tightness of cropland supply and limited reallocation possibilities among crops, and
- foreign trade policies that cut off supplies from international markets thus exaggerating international market supply tightness.

The study authors conclude that, until international stocks are able to rebuild to more traditional levels, agricultural commodity markets are likely to remain vulnerable to sharp price movements. Demand growth is expected to slow as corn biofuels mandates are reached and as China builds its soybean stock levels. However, in the absence of yields well above trend, it appears that the tight world stocks for corn, soybeans, and other feedstuffs, cannot be overcome in a single crop year but will be likely to persist for several years.

Issues for Congress

U.S. and global feed costs have moved sharply higher since 2007, in part due to their linkage (via biofuels policy) to energy markets, but also as a result of a tightening resource base, international market conditions, macroeconomic factors, and weather events that have contributed to a series of years where feed demand has outpaced supply. For the U.S. livestock and poultry sector—where feed costs account for 50% to 80% of cash operating expenses—this poses significant consequences including declining profit margins, increased financial pressure, and the prospects for substantial near-term liquidation of herds or flocks. Smaller future herds or flocks will ultimately translate into higher retail prices for beef, pork, poultry meat, eggs, and dairy products.

The current prolonged drought in the Southwest and Southeast has only aggravated the persistent tightening of feed supplies. Apart from buying expensive commercial feeds or liquidating herd and flock numbers, there appear to be few options in the near-term for addressing the situation. Under certain conditions—including a severe drought—ranchers and livestock producers are permitted by USDA to cut hay and graze livestock on land set aside under the Conservation Reserve Program (CRP).[25] However, this may offer little relief if the CRP lands are located within a drought zone.

The 2008 farm bill authorized three new livestock assistance programs.[26] The Livestock Forage Disaster Program (LFP) assists ranchers who graze livestock on drought-affected pastureland or grazing land.[27] The Livestock Indemnity Program (LIP) compensates ranchers at a rate of 75% of market value for livestock mortality caused by a disaster. The Emergency Assistance for Livestock, Honey Bees, and Farm-Raised Fish Program (ELAP) compensates producers for disaster losses not covered under other disaster programs. However, all three of these programs expire on September 30, 2011, and reauthorization might be difficult since they have no baseline funding beyond FY2011.

Perhaps the most effective measures for dealing with a drought and/or sharply higher feed costs are precautionary ones taken in the spring in advance of the onset of severe summer weather conditions when both federal risk management tools and commercial hedging operations can help mitigate any potential economic losses. For example, USDA offers limited livestock programs under the federal crop insurance program.[28] Relatively new or pilot programs protect livestock and dairy producers from loss of gross margin or price declines. In addition, several index-based policies are available to insure against losses on pasture, rangeland, and forage.[29] The U.S. dairy industry in particular has been concerned about the increasing price volatility of commercial feed markets and has been exploring the potential for developing a federal dairy program that focuses on stronger output-to-input margin protection.[30] In the longer term, federal policies closely associated with the decline in feed availability and high commodity prices are both biofuels policy and land-use policy (particularly the CRP).

For major program crop producers, high market prices—at levels above commodity loan rates and target prices—have resulted in a sharp decline in payments under the price contingent marketing loan and counter-cyclical payment programs. If commodity prices are sustained at such high levels, price contingent farm programs are likely to lose much or possibly all of their long- run baseline funding. These and other issues related to livestock feed availability are likely to be of growing interest to Congress as the House and Senate Agriculture Committees monitor the financial health and well-being of the U.S. livestock and poultry sectors as well as any future ramifications for retail food price inflation.

In addition, Congress, along with energy and agricultural market participants, is expected to closely follow any further acceleration of unanticipated side-effects on the U.S. livestock sector from continued corn use for ethanol production. Finally, feed availability and associated market price volatility are likely to play a significant role in the next farm bill debate, as the current omnibus farm bill expires in 2012. The decline in price contingent program payments under high commodity prices coupled with the increased price volatility of recent years is likely to refocus interest on federal support for risk management tools to help agricultural producers manage the unpredictability of the marketplace.

APPENDIX. OVERVIEW OF MAJOR FEEDSTUFFS

Primary Feedstuff Sources

The three main nutrient categories—roughage, energy, and protein—can be obtained from several sources.

Feed Grains—The Main Energy Source

Feed grains—corn, sorghum, barley, oats, and feed wheat—account for about 70% of all feedstuff (energy and protein alike) consumed by livestock.[31] The high starch levels of feed grains supply most of the energy needed for animal and poultry body maintenance and production of meat, milk, and eggs. Over the past decade, poultry (33%) and hogs (27%) have been the dominant users of feed grains compared with cattle on feed (24%) and dairy cows (11%).[32]

Corn: The King of Feed Grains

Corn is the world's most important feed. It is the base ingredient for most dairy, beef, swine, and poultry feeds. Corn accounts for 92% of all U.S. feed grains produced, and 65% of all U.S. feedstuffs (both energy and protein) consumed.[33] Corn's nutritional advantage includes high carbohydrate and fat contents, and low fiber content. However, corn is deficient in several amino acids, calcium, and most vitamins and must, therefore, be used in coordination with other specific feedstuffs to achieve a balanced diet for livestock and poultry.

The United States is the world's largest producer of corn. The U.S. corn crop is perennially the largest single crop of any kind grown by any country in the world. Since 2005, U.S. corn plantings have averaged over 86 million acres, with the majority of the crop grown in the Corn Belt (Figure A-1). The United States is also the world's leading consumer (38% of world use) and exporter (60% of global exports) of corn.[34] Since 2001, the United States has exported approximately 18% of its corn crop (about 2 billion bushels annually) to other countries. U.S. corn is prized in international markets for its nutritional profile and superior palatability. For example, egg producers in the Middle East value corn over other feedstuffs for its enhanced yellowing of egg yolks which adds value in retail markets.

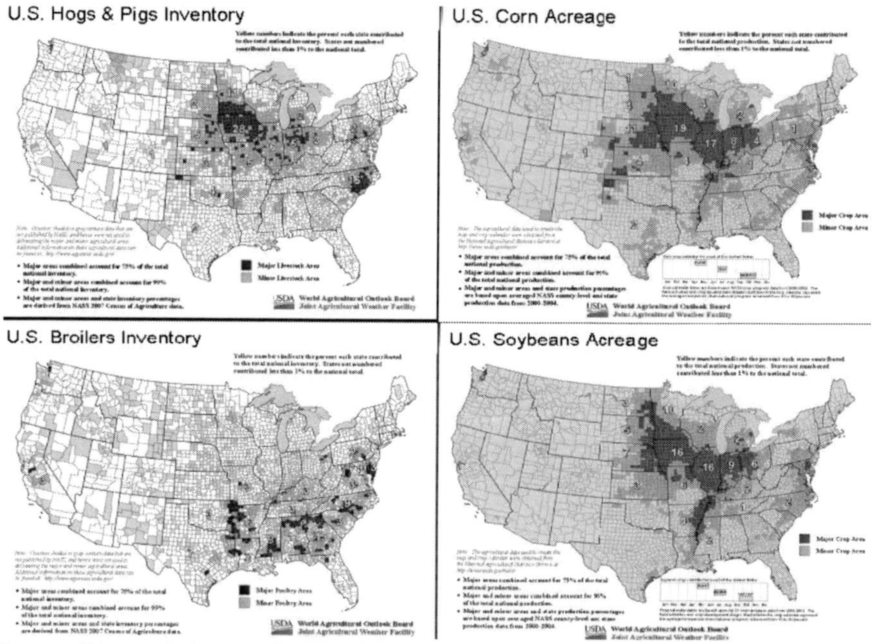

Source: Livestock inventories are for 2007, corn and soybean data are averages for 2000-04, WAOB, USDA.

Figure A-1. U.S. Hogs and Pigs and Broilers Inventories, Corn and Soybean Production.

In addition to its role as the world's most important feed grain, corn is also processed into a multitude of food and industrial products including starch, sweeteners, corn oil, beverage and industrial alcohol, and fuel ethanol. All of these alternate uses compete in the marketplace with livestock and poultry feeders for existing corn supplies. Traditionally most of the U.S. corn crop has been used as the main energy ingredient in livestock feed. However, since 2004 an increasing share of the U.S. corn crop has been consumed by the ethanol industry. In 1999 less than 6% of the U.S. corn crop was used for ethanol production and feed byproducts; this share has grown to over 40% in 2010. In 2011, USDA forecasts that the volume of corn used for ethanol will exceed corn fed to livestock for the first time in history.[35] This new demand source has put additional upward pressure on U.S. corn prices in recent years.

Feed demand for corn varies seasonally—its use is traditionally highest in the first (September- November) and second (December-February) quarters after harvest when supplies are most plentiful. By March, most pastures and ranges are flush with new growth and cattle ranchers take advantage by shifting some of their animals onto the green forages. During the summer months hay and other roughages are more abundant and compete in feed rations of ruminants. Corn use also varies substantially by region. Surplus corn-producing regions include the Corn Belt, Lake States, and Northern Plains (Figure A-1). All other regions have deficits.

Sorghum, Barley, and Oats: Declining Importance as Feed Grains

Sorghum, barley, and oats are the three remaining major feed grains although they pale in significance when compared with corn. Grain sorghum and barley have about 95% of the feed value of corn, while oats has about 90% of corn's feed value. In the past decade sorghum has accounted for about 3% of feed grain use compared with 1% shares for both oats and barley. The tremendous growth in corn yields (and significant improvements in its genetic adaptation to varying agro-climatic settings) relative to other field crops has allowed corn to expand into nontraditional areas at the expense of minor feed grains, wheat, and oilseed crops. As a result, corn's planted acreage and output has been steadily increasing since the early 1980s, while sorghum, barley, and oats have been on the decline (Figure 11).

Sorghum is a drought- and heat-resistant crop grown primarily in the Central and Southern Plains regions where the climate is often impractical for non-irrigated corn (Figure A-2). Most sorghum is consumed as a feed grain in Southwestern cattle feedlots. Sorghum also rivals corn as a potential ethanol feedstock and is used to produce biofuels in a few ethanol plants located close to sorghum growing regions. The United States is the world's leading sorghum exporter, shipping about 46% of annual production (or about 188 million bushels), primarily to Japan and Mexico.[36]

Barley production is concentrated in the Northern Plains (Figure A-2). Its most valued use is as a source of malt for the brewing industry. However, poor barley quality (usually related to growing conditions or excessive harvest-time moisture) and relative feed grain price differentials can lead to substantial barley feeding to cattle and sheep.

Oats production (and its use for feed) in the United States has declined steadily since the 1950s for a variety of reasons including its lower feed value and price relative to other feed grains. Since the mid-1980s the United States has become the world's largest importer of oats, primarily for use as horse feed. During the past decade, the United States has imported nearly 48% (or about 98 million bushels) of its annual oats consumption.

Feed Wheat: Varies with Quality and Market Conditions

Wheat is generally grown as a food crop, but may also be used as a livestock feed since its feed value is comparable to that of corn. Wheat's traditionally higher price relative to corn is a major factor in determining the extent to which wheat is used as a livestock feed. However, weather- related damage to wheat, particularly excessive moisture at harvest time, can render the grain unsuitable for milling uses and leave feeding as the only option. During the past decade, about 7% of the annual U.S. wheat crop has been fed to livestock.[37] However, wheat feeding can vary substantially from year to year depending on crop size and its price relative to corn. For example, in 1990 about 20% of the crop was fed to livestock, while in 2007 less than 1% was used as feed. In general, supply tightness of traditional feed grains (whether from drought or heavy export demand) or abundant wheat supplies can price wheat into greater feed use.

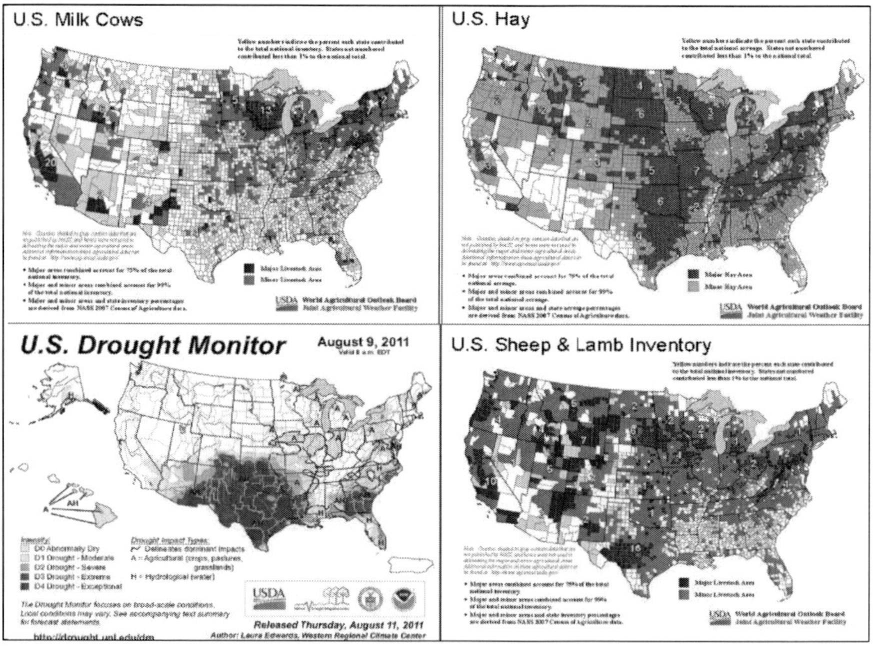

Source: Wheat, sorghum, and barley production data are averages for 2000-04, WAOB, USDA.

Figure A-2. Winter Wheat, Sorghum, and Barley Production, and U.S. Drought Monitor.

Wheat feeding is most common in corn-deficit areas such as the Southern Plains, where little corn and much hard red winter wheat are grown (**Figure A-2**). As a result, beef cattle on Southwestern feedlots have been the traditional consumers of feed wheat, although poultry in Southeastern operations are also occasionally fed wheat.

Protein Meals

Protein supplements are needed to correct the protein deficiencies of an all-grain diet. However, protein meals are usually more expensive than feed grains. Most protein concentrates used in the United States are oilseed meals (61% share)—derived from the pulp residue left over from the crushing process used to extract vegetable oil from oilseeds, for example, soybeans, cottonseed, rapeseed or canola, peanuts, and minor oilseeds (sunflowers, linseed, sesame, and mustard). Additional protein meals are obtained from byproducts from processing grains (34% share including DGS), animals (4%), and marine species (< 1%).

Soybeans are the major source of protein meal in both the United States and the rest of the world. Globally, soybean meal (or soymeal) represents between 65% to 70% of all protein meals produced, while in the United States soymeal represents 85% of all oilseed meals and 53% of all protein meals fed to U.S. livestock and poultry.

The rapid growth of poultry meat production and its higher feed efficiency relative to other species has changed the nature of concentrated feed use in the United States. Poultry require less feed to produce a pound of meat than other animals (Table A-1). However, poultry production required increasing quantities of protein as the industry expanded significantly since the 1970s. Poultry's consumption share of high-protein feed ingredients in the United States has risen from a 40% share in the early 1970s to 57% during the past decade. Hogs' share of high-protein feeds has held fairly steady at about 22% over the past 40 years, while dairy and beef cattle have seen their shares drop from around 12% to 13% in the early 1970s to 7% for dairy and 9% for beef cattle in recent years.

Oilseed Meals

Most protein meals are derived from the pulp residue left over from the crushing process used to extract vegetable oil from oilseeds—for example, soybeans, cottonseed, rapeseed or canola, peanuts, and minor oilseeds (sunflowers, linseed, sesame, and mustard). Soybeans are the major source of protein meal in both the United States and the rest of the world. Globally, soybean meal (or soymeal) represents between 65% to 70% of all protein meals produced, while in the United States soymeal represents 85% of all oilseed meals and 53% of all protein meals fed to U.S. livestock and poultry.

Soymeal comprises the largest part of a soybean—about 75% of the volume and 60% of the processed value. In contrast, soybean oil or soyoil accounts for about 19% of volume and 39% of the processed value.[38] The value share can vary from year to year such that in some years the price of oil is high enough relative to meal that it commands a majority of the total processed oil and meal value. Both the oil and meal have positive market characteristics that make them very competitive with other vegetable oils and meals.

Soybeans are a high-yielding, broad-leaf leguminous oilseed plant that fits well in rotation with corn to control plant pests and diseases. As a result, the Corn Belt has the largest concentration of soybean acreage in the United States (Figure A-1), although genetic improvements have helped to expand U.S. soybean production into the Northern Plains. Soybeans require less water than corn and, as a result, tend to have a larger growing zone that also includes the Southeast, Mid- Atlantic, and Central Plains.

Other oilseed crops tend to be both lower yielding and less profitable per acre to produce than soybeans. As a result, most other oilseeds are relegated to agro-climatic zones where soybeans do not grow well. The oil-to-meal ratio varies considerably across oilseeds. For example, rapeseed is about 40% oil and 60% meal. But most of the meals derived from other oilseeds (in particular, cottonseed, canola, and sunflower) are higher in fiber and tougher to digest than soymeal. This means that they have a lower digestible energy content for swine and poultry. As a result, most other oilseed meals are used predominantly by the cattle industry. Other oilseed meals account for about 9% of total protein meals fed in the United States.

Animal Meals

Many feeds make use of the byproducts generated in the meat, poultry, milk, and fish processing industries. Organic materials that would otherwise go to waste can be rendered into valuable, high-quality protein products that are useful as supplements to other energy and protein concentrates to achieve a balance of essential vitamins and amino acids.

Meatpacking byproducts include meat, bones, blood, and animal fat which can be used to produce meat and bone meal and blood meal. Their use represents about 4% of total protein meal use in the United States. However, stringent feed-use restrictions related to potential transmission of bovine spongiform encephalopathy (B SE) have been placed on certain animal byproducts which limit both their current marketability and their growth potential.[39]

Roughages

Roughage (i.e., cellulose-based) feedstuffs are vegetative plants that are bulky, high in crude fiber, and low in digestible nutrients. The major groups of roughage include pasture and range plants, hays, and silages. *Dry roughage* is obtained from cutting and harvesting *hay* (i.e., legumes, grasses, or grain crops cut at an early growth stage to maximize nutrient yield) grown on rotated cropland or pasture. *Green forages* include *silage* which is a type of forage (often based on chopped hay or grain crops) that is fermented in a silo, or *grazing* the crops (i.e., wheat and rye), grasses, or wild plants on cultivated pasture or rangeland. An advantage of silage is that it preserves a succulent forage for winter feeding.

Roughages are most suitable for use by ruminants—for example, cattle, sheep, and goats—which have micro-organisms in their digestive tracts that secrete enzymes which break down the cellulose in plant fiber. Ruminants actually require a high level of fiber in their diets to maintain a rumen environment favorable for the synthesis of nutrients. Insufficient fiber in a ruminant's diet can lead to acute digestive disorders. Roughages supply as much as 55% to 60% of the annual feed consumption by dairy and beef cows and sheep; however, roughages do not fully replace feed grains and protein meals in a commercial feeding operation because they cannot supply the high levels of net energy and protein needed to produce rapid weight gains and high milk yields.

Since 2001, dairy cows have consumed about 18% of U.S. roughage production, while beef cattle operations account for about 70% of all roughage in the United States.[40]

Corn and sorghum are the most important silage crops because they have the highest per-acre yields of digestible nutrients. However, when corn and sorghum are used for silage, the entire plant—stalk, leaves, and grain—is harvested, chopped, and often mixed with other silage. Most silage is produced and used on-farm. Marketing silage is uncommon because it is costly to transport and loses its quality quickly once removed from the silo.

Pasture and Range Flexibility

Livestock is the residual claimant on agricultural land. If any other farm activity can be successfully undertaken on a piece of land, it usually takes priority over livestock grazing. As a result, pastures (or participation in the Conservation Reserve Program[41]) may be the best use of marginal land unable to profitably grow cash crops. (For example, compare the location of cattle, dairy, and sheep populations of Figure A-3 with the location of major field crops in Figure A-1 and Figure A-2.) In primary crop production areas such as the Corn Belt, pastures are increasingly being retained only for the use of dry dairy cows, calves and replacement heifers, and breeding beef cows or sheep. This pattern has accelerated over the past several years in response to higher crop prices.

Ranges—located predominantly in Western States—are less productive than cultivated pastures because they usually have lower rainfall, poorer soils, shorter growing seasons, and steep topography. Both pastures and ranges are more productive in the spring and early summer when young, leafy plants are higher in moisture and lower in fiber, making it more palatable and digestible to livestock than later summer stages when it is coarser and stemmier. The period of pasture and range fertility is closely linked to weather conditions and can be extended into the summer or curtailed in early spring based on rainfall patterns.

Having access to good quality pasture, rangeland, or small grain crops for grazing provides substantial flexibility for ranchers to reduce their costs by purchasing fewer commercial feed concentrates, while minimizing their operating expenses for growing, harvesting, feeding, and storing grains or silage. When feed grain prices increase, many producers will graze cattle longer before placing them into feedlots at heavier weights to save on finishing costs, provided that there is sufficient forage. The ongoing drought in the Southern Plains and Southeast has severely limited grazing opportunities for cattlemen.

Using Feed Grains

Feed grains may be purchased commercially or produced and used directly on the farm.

Commercial Feeds

Commercial feed concentrates offer the advantage of ready-to-eat form and precise nutrient composition, but are generally more expensive than home-grown feeds. Different milling procedures are used to prepare feed grains for animal consumption. Grinding the grain reduces the animal's expenditure of energy in chewing and exposes a greater surface for improved digestion. The desired fineness of the grind depends on the type of livestock and other components being blended in the ration. Cracking grain is sufficient for cattle; fine grinds are better for poultry. Grinding also assists in handling and pelleting (compressing into pellets), but adds cost to every bushel of grain.

On-Farm Feeding

Farmers that produce both livestock and feedstuffs (whether grains, hay, or silage), will normally feed with their own-farm-produced feedstuffs before turning to commercial feed grain markets. Substantial on-farm feeding of own-produced feedstuffs occurs throughout the Corn Belt, Lake States, and Northern Plains where surplus production of feedstuffs exists. In feed-deficit regions, commercial feed markets and feed-pasture decisions tend to play a much larger role.

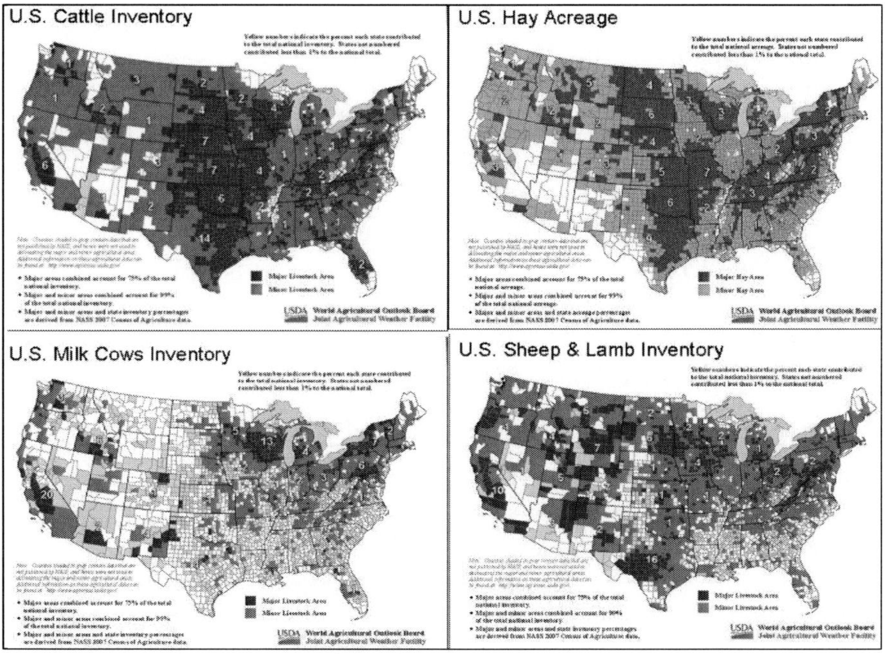

Source: Data are for 2007 from WAOB, USDA.

Figure A-3. U.S. Cattle, Milk Cows, and Sheep and Lambs Inventories and Hay Acreage.

The amount of on-farm feeding from own-production can also be influenced by the quality of the crop. When late planting delays crop maturity, or an early frost or substantial harvest-time rain occurs, farmers may be confronted with the choice of expensive crop drying or on-farm storage of high-moisture (20% to 30% moisture) grain. Under proper storage conditions, high-moisture grain is a high-valued feed. However, because of its high-moisture content, it is not readily marketed off the farm due to rapid spoilage. As a result, a commitment to store on-farm high- moisture grain is also a commitment to feed it on farm rather than sell it in commercial markets.

Feed Conversion Ratio (FCR)

In animal husbandry, the feed conversion ratio (FCR) is a measure of an animal's efficiency in converting feed mass into increased body mass. Animals that have a low FCR are considered efficient users of feed.

Grain Milling Byproducts

Feed grain byproducts originate primarily from milling grains and include corn, wheat, and rice mill feeds and a variety of miscellaneous feed byproducts such as hominy feed, oat mill feeds, and screenings. This group of other feed grain byproducts is fairly small, but not insignificant, accounting for about 4% total feed concentrates.

Table A-1. Feed Conversion Ratio (FCR) for Major Livestock and Bird Species

FCR is the ratio, by weight, of feed needed to produce one pound of meat.

Species	Feed Conversion Ratio
Cattle	6
Sheep	5 to 6
Lambs	3 to 4
Hogs	3 to 3.2
Broilers	2
Turkey	2

Source: Compiled by CRS from various sources.

Notes: The actual FCR will vary widely based on the quality of feedstuffs, climate (extreme temperatures may require more calories for maintenance), animal activity, and other conditions.

Corn milling byproducts have grown in importance with the rapid growth of the corn-based U.S. ethanol industry. Two primary corn milling processes, wet and dry milling, are used to produce fuel ethanol. Both processes also produce several categories of important byproducts that are attractive animal feed alternatives, particularly as both energy and protein concentrates.

Corn wet-milling is designed to extract starch, high-fructose corn syrup (HFCS), glucose and dextrose, and ethanol from corn. Wet-milling byproducts include corn gluten feed (CGF) and corn gluten meal (CGM).[42] Corn dry-milling is used to produce distilled liquor and ethanol (fuel alcohol). Dry-milling byproducts include distillers grains and solubles (DGS).[43]

Distillers Grains and Solubles (DGS)

Since mid-2005 all new corn milling plants built in the United States have been corn dry milling plants such that dry milling production of ethanol and its byproducts now dominates the U.S. ethanol industry. As a result, production of distillers grains and solubles (DGS)—the major co- product of the corn-starch ethanol industry—has expanded very rapidly in the last six years with the explosive growth in corn processing for ethanol (Figure 5), rising from about 1 million metric tons (mmt) in the 1999/2000 corn marketing year (Sept. -Aug.) to 33.2 mmt in 2010/2011.

DGS is a valuable feed alternative. DGS is a medium-protein (approximately 30% protein content) feed ingredient that replaces corn, soybean meal, and other sources of phosphorus and calcium. In addition, its percentages of fiber, protein, oil, and other nutrients are concentrated to almost three times the percentages found in corn. Initially, the U.S. livestock and poultry industry was slow to adopt DGS into feed rations and instead required a time lag in order to develop systems for efficiently transporting, handling, and feeding this product.[44] However, feeding of DGS is now widespread in the United States, especially in the beef cattle and dairy industries. When DGS is sold to distant domestic feed markets or for export, it must first incur a drying cost and then is sold as dried distillers' grains and solubles (DDGS). In the last few years, DDGS exports have grown rapidly as foreign livestock and poultry producers became familiar with its potential cost savings and ways of effectively using it as a replacement for corn and soymeal.

Where ethanol plants are located close to feedlots, a portion of the DGS output typically is sold as wet or partially-dried DGS, and is moved directly to cattle feedlots. Wet and partially dried distillers grains have limited storage life, since they contain 70%-75% and 50%-55% moisture, respectively. As a result, wet DGS is sold at a discount to costlier DDGS giving a cost advantage to livestock feeders located within close proximity of a dry-mill ethanol plant (Figure A-4).

DDGS is best suited for feeding to ruminant animals (e.g., cattle, sheep, goats). The digestive system of these animals allows them to convert DGS fiber into energy while at the same time using the protein and other nutrients. Use of DDGS in hog and poultry rations is somewhat more limited, in part because of its fiber content. When DDGS is fed to these species, supplementation of some amino acids also is needed. Pork producers report that feeding high levels of DDGS during the feeding period can cause soft bellies that in turn cause problems in bacon production.

Recommended maximum DDGS percentages of the ration for swine and poultry are considerably lower than for cattle because of differences in their digestive systems (Table A-2). As a result, beef cattle are the largest market for DDGS by a considerable margin, followed by its use for dairy cows, hogs, and poultry in that order. Replacement ratios in substituting DDGS for corn and soybean meal also vary with the species of animal or bird being fed.

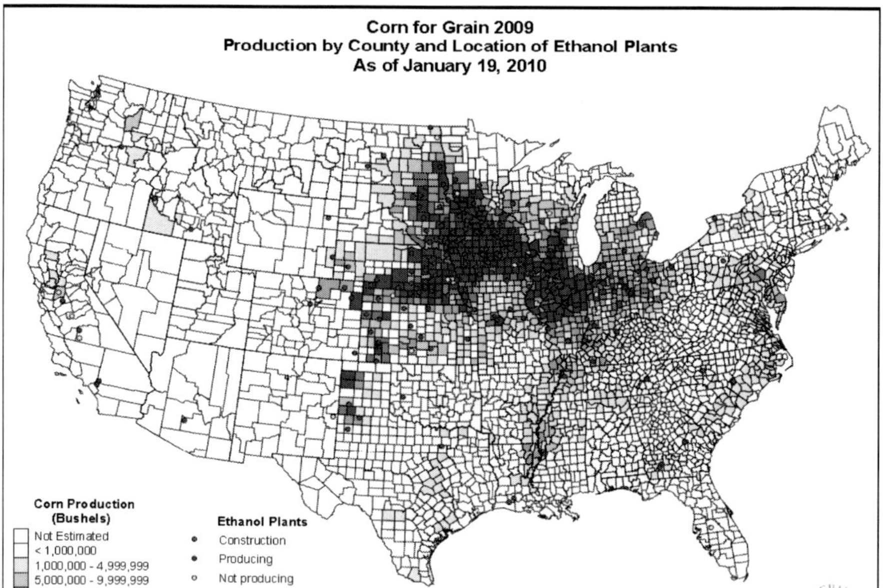

Source: NASS, USDA.

Figure A-4. Comparison of U.S. Ethanol Plant Locations with U.S. Corn Production.

Table A-2. Recommended DGS Usage Rates in Livestock and Poultry Rations

Species	Share of Ration[a]	One pound of DGS replacesb	
		Corn	Soymeal
Beef Cattle	over 50%	1.1 lbs.	—[c]
Dairy Cows	up to 40%	0.6 lbs.	0.6 lbs.
Hogs	10% to 15%	0.8 lbs.	0.2 lbs.
Poultry	4% to 7%	0.6 lbs.	0.6 lbs.

Sources:

a. Ration-share estimates are compiled by CRS from various sources; however, it is noteworthy that there does not yet appear to be any consensus on the exact ranges of DDGS inclusion in rations for the various animal and bird species.

b. Feed-replacement estimates are from Robert Wisner, "Distillers' Grain Price Relationships and Export Developments," *AgMRC Renewable Energy and Climate Newsletter*, AgMRC, December 2010.

c. Very little soymeal is used in typical U.S. beef cattle feeding programs.

End Notes

1 For a description of the various major feed sources, see the Appendix (*Overview of Major Feedstuffs*) to this report.

2 The material in this report draws heavily from *Animal Feeds Compendium*, AER No. 656, Mark Ash, Economic Research Service (ERS), USDA, May 1992.

3 See the Appendix of this report (*Overview of Major Feedstuffs*) for a discussion of the various feed types and sources.

4 For a similar discussion of the market forces behind high commodity prices and declining feed availability (but with more detail and a slightly different emphasis) see Philip Abbott, Christopher Hurt, and Wallace Tyner, *What's Driving Food Prices in 2011?*, Farm Foundation Issue Report, July 2011; hereafter referred to as Abbott, Hurt, and Tyner, *What is Driving Food Prices in 2011*, Farm Foundation Report, July 2011.

5 CRS Report RL3 3204, *Price Determination in Agricultural Commodity Markets: A Primer*.

6 WAOB, USDA, *WASDE*, July 12, 2011.

7 The 2004/2005 corn marketing year was a year of relatively high stocks-to-use (20%), but it was also the year that preceded the "take-off" of the U.S. ethanol industry. For more details, see CRS Report R41282, *Agriculture-Based Biofuels: Overview and Emerging Issues*.

8 Sechler, Bob. "U.S. Lament: How Dry We Are—Forecasters Say Drought to Persist; Crops, Cattle Affected," *The Wall Street Journal*, July 8, 2011.

9 Ibid., David Hutchins, cattle broker with Amarillo Brokerage, Amarillo, Texas.

10 CME Group, "Persistent Drought Threatens Wheat Planting in Southern Plains," *Commodity News for Tomorrow*, (www.cmegroup.com), August 2, 2011.

11 CRS projection based on data from the "Table 10.3 Fuel Ethanol Overview," *Monthly Energy Review*, U.S. Energy Information Administration (EIA), U.S. Dept. of Energy, June 2011. Note that Figure 6 compares total-use shares (excluding use for stocks); whereas Figure 5 expresses the usage rates as shares of annual crop production.

[12] Abbott, Hurt, and Tyner argue that the adjustment factor is 27.6% per bushel of corn, calculated as 18 lbs. of DGS per 56 lbs. of corn corrected by DGS' lower feed value—86% of corn; see Philip Abbott, Christopher Hurt, and Wallace Tyner, *What's Driving Food Prices in 2011?*, Farm Foundation Issue Report, July 2011.

[13] WAOB, USDA, *WASDE*, August 11, 2011.

[14] See the Appendix to this report for a detailed discussion of DGS use as an animal feed.

[15] See FAPRI, "New Challenges in Agricultural Modeling: Relating Energy and Farm Commodity Prices," FAPRI-MU Report #05-11, Food and Agricultural Policy Research Institute, May 2011.

[16] CRS Report R40155, *Renewable Fuel Standard (RFS): Overview and Issues.*

[17] For example, see *U.S. Baseline Briefing Book*, Food and Agricultural Policy Research Institute (FAPRI), FAPRI-MU Report #02-11, March 2011.

[18] Data is for 1978-2009 from "China: GDP Growth (Annual %)," *World Development Indicators*, World Bank.

[19] "China In-Country Analysis," *Ag Perspectives*, World Perspectives, Inc., July 11, 2011.

[20] Abbott, Hurt, and Tyner, *What's Driving Food Prices in 2011*, Farm Foundation Report, July 2011.

[21] "China Buys Nearly 4 Mln Tons U.S. Corn—Executive," *CME Daily Grain Newsletter*, CME Group, July 12, 2011.

[22] *USDA Agricultural Projections to 2020*, OCE-201 1-1, WAOB, USDA, February 2011.

[23] See CRS Report RS21613, *Conservation Reserve Program: Status and Current Issues.*

[24] Abbott, Hurt, and Tyner, *What is Driving Food Prices in 2011*, Farm Foundation Report, July 2011.

[25] See "Emergency Haying and Grazing," Conservation Programs, Farm Service Agency (FSA), USDA, at http://www.apfo.usda.gov/FSA/webapp?area=home&subject=copr&topic=crp-eg.

[26] See CRS Report RS21212, *Agricultural Disaster Assistance.*

[27] See USDA fact sheet at http://www.fsa.usda.gov/FSA/webapp?area=home&subject=diap&topic=lfp. Also see FSA, USDA, "Livestock Forage Disaster Program and Emergency Assistance for Livestock, Honeybees, and Farm-Raised Fish; Supplemental Agricultural Disaster Assistance," 74 Federal Register 46665-46683, Sept. 11, 2009.

[28] See CRS Report R40532, *Federal Crop Insurance: Background and Issues*, and CRS Report R41 854, *Risk Management Tools for Dairy Farmers.*

[29] See Risk Management Agency (RMA), USDA, "Pasture, Rangeland, and Forage," *Crop Policies and Pilots*, available at http://www.rma.usda.gov/policies/pasturerangeforage/.

[30] See CRS Report R41 141, *Previewing Dairy Policy Options for the Next Farm Bill.*

[31] Note that USDA data exclude distillers' dried grains and solubles (DDGS) for lack of official data. However, CRS has used unofficial data sources to incorporate the DDGS feed estimates into the USDA data to produce a more complete estimate. USDA data are from "Table 29—Processed feeds: Quantities fed and feed per grain-consuming animal unit," *Feed Grains Database: Yearbook Tables*, ERS, USDA, June 9, 2011. Unofficial DDGS data sources are described in *Market Issues and Prospects for U.S. Distillers' Grains: Supply, Use, and Price Relationships*, FDS-10k01, Linwood Hoffman and Allen Baker, ERS, USDA, December 2010.

[32] Calculations by CRS based on data from "Table 30—Indexes of Feed Consuming Animal Units (millions)," *Feed Grains Database: Yearbook Tables*, ERS, USDA, June 9, 2011.

[33] Based on CRS-adjusted USDA processed feed estimates as described in the earlier footnote.

[34] *WASDE*, WAOB, USDA, July 12, 2011.

[35] WAOB, USDA, *WASDE*, July 12, 2011.

[36] Foreign Agricultural Service (FAS), USDA, *Grain: World Markets and Trade*, June 2011.

[37] FAS, USDA, Production, Supply and Distribution Online, July 12, 2011, at http://www.fas.usda.gov/psdonline/.

[38] "Table 9—Soybeans: Monthly value of products per bushel of soybeans processed, and spot price spread, U.S., 1990/91-2009/10," *Oil Crops Yearbook Database*, ERS, USDA, last updated March 2011.

[39] For more information see CRS Report RL32199, *Bovine Spongiform Encephalopathy (BSE, or "Mad Cow Disease"): Current and Proposed Safeguards*.

[40] ERS, USDA, "Table 30—Indexes of feed consuming animal units," *Feed Grains Database: Yearbook Tables*, June 9, 2011.

[41] The Conservation Reserve Program (CRP) is a federal program, enacted in the 1985 farm bill, that provides payments to farmers to take highly erodible or environmentally sensitive cropland out of production for 10 years or more.

[42] A 56-pound bushel of corn yields about 12.5 pounds of CGF and 2.5 pounds of CGM. Linwood Hoffman and Allen Baker, *Market Issues and Prospects for U.S. Distillers' Grains: Supply, Use, and Price Relationships*, FDS-10k-01, ERS, USDA, December 2010.

[43] A bushel of corn processed into ethanol by dry mills produces approximately 17.5 pounds of DGS, along with carbon dioxide, and 2.7 gallons of ethanol. Robert Wisner, "Distillers' Grain Price Relationships and Export Developments," *AgMRC Renewable Energy and Climate Newsletter*, Agricultural Marketing Resource Center (AgMRC), December 2010; at http://www.agmrc.org/.

[44] Ibid.

In: Feed Market Dynamics …
Editors A. Stewart and K. Gray
ISBN: 978-1-61942-272-8

Chapter 2

ETHANOL CO-PRODUCT USE IN U.S. CATTLE FEEDING: LESSONS LEARNED AND CONSIDERATIONS*

Kenneth H. Mathews, Jr. and Michael J. McConnell

ABSTRACT

The byproducts of making ethanol, sweeteners, syrups, and oils used to be considered less valuable than the primary products. But the increased livestock-feed market for such byproducts in the past few years has switched that perception to one of the ethanol industry making grain-based "co-products" that have market value separate from the primary products. Co-products such as dried distiller's grains, corn gluten feed, corn gluten meal, corn oil, solubles, and brewer's grains have become economically viable components, along with traditional ingredients (such as corn, soybean meal, and urea), in feed rations. The co-products have limitations, such as variable moisture content, product availability, nutrient excesses or deficiencies, and nutrient variability. These limitations affect how they must be handled and stored and how much they cost feed buyers. Dried distiller's grains are more amenable to pelleting and other bulk-handling methods than other co-products, which gives them an advantage in international markets.

* This is an edited, reformatted and augmented edition of a United States Department of Agriculture publication, Report FDS-09D-01, dated April 2009.

ACKNOWLEDGMENTS

Thanks to the numerous reviewers and others who provided useful comments. Thanks especially to Don Blayney, Linwood Hoffman, Greg Pompelli (USDA, ERS), Harry Baumes (USDA, Office of Energy Policy and New Uses), John Lawrence (Iowa State University), Darrell Mark (University of Nebraska-Lincoln), Gerald Norton (USDA, World Agricultural Outlook Board), and Hosein Shapouri (USDA, Office of Energy Policy and New Uses) for their thoughtful reviews and helpful comments. Thanks also to Priscilla Smith and Wynnice Pointer-Napper for their editorial/design expertise.

INTRODUCTION

The expansion of the ethanol industry led to increased demand for corn and an increased supply of co-products from the ethanol production processes. Concurrently, high global commodity prices, attributed to increased global demand for food and feed, poor weather conditions in some major crop-producing areas, and the worldwide expansion of biofuel production, have detrimentally impacted the livestock feeding industry (Trostle, 2008). Co-products from ethanol, sweeteners, syrup, and grain-based oil production have been used as livestock feed for many years, but only recently have become a more widespread economic alternative for coping with high prices for feedstuffs such as corn, soybean meal, and urea (figure 1). Dairy cattle operators have the most experience of agricultural producers who use co-products. Dairy operators, surveyed by the National Agricultural Statistics Service (NASS) in 2007 indicated that, on average, they had fed co-products 9.2 years (USDA, NASS, 2007). Cattle feeders and cow-calf operators indicated they have used them, on average, for 5.1 years and 4.6 years, respectively. Co-product feeding is relatively new for hog operations, with survey respondents indicating they had used co-products 2.7 years on average.

Ethanol plants increasingly compete with livestock feeders for corn. However, ethanol production also results in co-products that can be substituted as sources of energy and protein in livestock feeding (table 1). The removal of starch through ethanol and sweetener production reduces the quantity of the feedstuff available in a bushel of corn or other grain. For example, a bushel of corn (56 pounds) that yields 2.6 to 2.8 gallons of ethanol reduces in the dry-mill process to about 17.5 pounds of dried distiller's grains with solubles containing 10-percent moisture. The same bushel in the wet-mill process reduces to 11-13 pounds of corn gluten feed, 2.6 pounds of corn gluten meal, and 1.6 pounds of corn oil (Biomass Research and Development Board, 2008; Westcott, 2007; Sevcik, 2006; Wisner, 2007; Shapouri, 2009).

Co-products have been supplied historically from industrial uses of corn and other grains, which have grown in recent years with increased ethanol production (table 2). While some corn usage for feed has been diverted to ethanol production, reducing feed uses of corn and other grains, the bulk of increased ethanol-driven corn usage has come from increased corn acreage, often a result of reallocating crop acreages. As a result, the total supply of grain-based feeds is expanding (figure 2). Thus, the increased availability of co-products has actually increased feed supplies (table 3). This has led to expanded interest in feeding co-products and periods of favorable relative prices have increased their substitution for corn.

Co-products made from grain-starch-based and corn-oil products include corn gluten feed (CGF), corn gluten meal (CGM), wet distiller's grains with solubles[1] (WDGS), dried distiller's grains without solubles (DDG), dried distiller's grains with solubles (DDGS), and solubles alone (condensed distiller's solubles (CDS)).

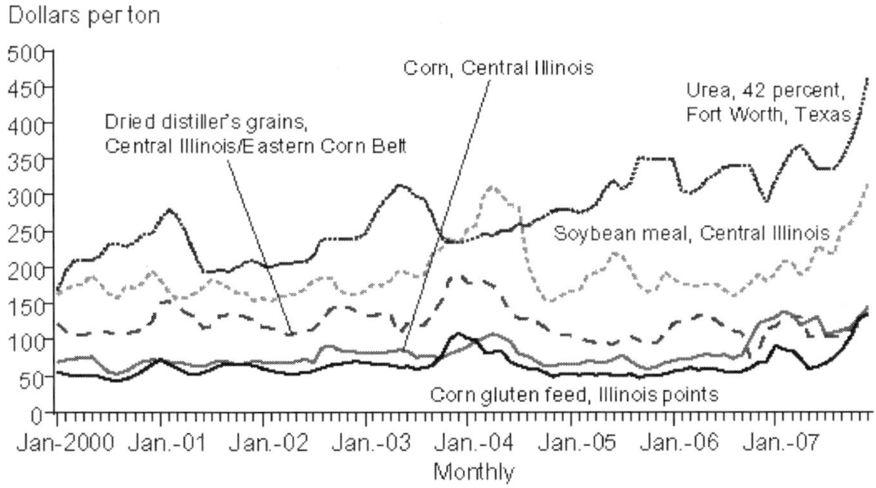

Source: USDA, ERS, Feed Grains Database, www.ers.usda.gov/Data/FeedGrains/.

Figure 1. U.S. prices for selected feedstuffs.

Table 1. Nutrient profiles of selected feedstuffs

	Dry matter	Crude protein	Total digestible nutrients	Lysine	Methionine	Sulfur
	Percent					
Corn	87	7.5	80	.24	.18	.08
Soybean meal	88	47.8	79	.71	.70	.43
Dried distiller's grains with solubles	92	27	82	.80	.51	.30
Wet distiller's grains with solubles	34.9	31	91			
Corn gluten feed	88	21	75	.60	.50	.16

Source: *Feedstuffs: 2009 Reference Issue and Buyers' Guide,* 2008.

Table 2. U.S. Ethanol Production

Year	*Million gallons*
1990	748
1991	866
1992	1,083
1993	1,154
1994	1,289
1995	1,358
1996	973
1997	1,288
1998	1,405
1999	1,465
2000	1,622
2001	1,765
2002	2,140
2003	2,804
2004	3,402
2005	3,904
2006	4,884
2007	6,521
2008	9,237

Source: U.S. Department of Energy, Energy Information Administration. http://www.tonto.eia.doe. gov/dnav/pet/pet_pnp_oxy_dc_nus_mbbl_m.htm/.

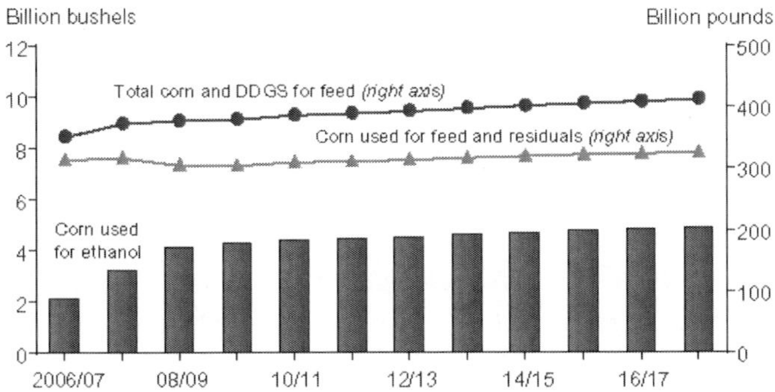

Note: DDGS = Dried distiller's grains with solubles.

Source: *USDA Agricultural Projections to 2018,* www.ers.usda.gov/publications/oce091/.

Figure 2. Total feed projections.

Table 3. Corn Provides Inputs Into Both Feed and Ethanol Production

Market year[1]	Production	Food, seed, and industrial use, minus fuel production	Fuel ethanol production	Feed and Residual use	Total domestic use
	Million bushels				
2001	9,502.58	1,340.42	705.95	5,864.27	7,910.63
2002	8,966.79	1,344.71	995.5	5,562.88	7,903.10
2003	10,087.29	1,369.56	1,167.55	5,793.02	8,330.13
2004	11,805.58	1,363.79	1,323.21	6,155.47	8,842.47
2005	11,112.19	1,378.39	1,603.32	6,152.28	9,133.99
2006	10,531.12	1,370.75	2,119.50	5,591.00	9,081.25
2007	13,037.88	1,337.29	3,026.13	5,938.13	10,301.55
2008	12,101.24	1,300.00[2]	3,600.00[2]	5,300.00[2]	10,200.00[2]

[1] Marketing year is September to August.

[2] Projected.

Source: USDA, ERS, Feed Grains Database, www.ers.usda.gov/Data/FeedGrains/.

Ethanol, Sweetener, and Starch Production Process

The co-products that are the focus of this chapter are created in the production processes of wet-mill and dry-mill corn plants. Corn and, in some cases, other grains are used as feedstocks in these mills to make primarily starch- related products (ethanol, sweeteners, and syrups) as well as oils. The wet- and dry-milling processes produce primary products and co-products with different characteristics.

Wet-mill plants use an intensive process to separate different parts of a corn kernel to produce syrups, ethanol, corn-starch-based plastics, oil, and other products. In the steeping process, corn kernels are softened in fresh water before being separated into bran, starch, protein, germ, and soluble components. Each component then goes through its own manufacturing process to create additional products. For example, the starch components are used to make ethanol. Bran and the fibrous remnants from the steeping process, distiller's solubles, are combined to make corn gluten feed.

Wet-mill plants are not as common as dry-mill plants, being more capital-intensive and requiring more investment per bushel of corn. However, wet-mill plants' revenue streams are more diversified because they create a variety of high-value products from a single kernel. There have been no new wet-mill plants built in the United States in the past 10 years, but a new one is being built in Iowa and production capacities of existing plants are expanding (Lawrence, 2009).

Dry-mill plants, designed to produce one product, corn-based-ethanol, have been at the center of the expansion of ethanol production in the United States over the past few years. The dry-mill process is less complex than the wet-mill process. Dry milling cooks and ferments ground corn, using enzymes and yeast to produce ethanol and carbon dioxide. Once the starch is converted to an alcohol, the leftover stillage is turned into distiller's grains. Dry-mill plants are generally smaller and require less capital investment per bushel of corn than wet-mill plants. However, some dry-mill plants are extracting more co-products through investment in fractionation and separation technologies (Lawrence, 2009). Although most are located in the Corn Belt (Illinois, Iowa, Minnesota, Nebraska, and South Dakota), plants are also located, planned, or under construction in other States (California, Georgia, New York, Texas, and Washington).

The value that co-products add to profit margins has become critical in maintaining processing plants' economic viability. When ethanol prices and profit margins for ethanol production were high, co-products were not as important as a source of revenue for ethanol plants. More recently, as ethanol prices have fallen and co-product demand has risen from livestock producers, co-product prices have been boosted to levels that, at times, exceeded the value of corn for feeding.

EXPERIENCE AND EDUCATION ENHANCE NUTRIENT MANAGEMENT

Co-products possess generally attractive nutrient profiles that render them useful as ingredients in feed rations for livestock, poultry, pets, and other animals (see table 1). Energy contents are on a par with feed grains, and co-product protein content usually falls between the lower protein content of grains and the higher content of oilseed meals. Based on USDA's baseline projections, co-products could increase the total supply of energy and protein available for livestock feeding (figures 3 and 4). Among the livestock species, animals with a rumen compartment in their stomachs are best able to utilize co-products in feed rations.[2] According to a National Agricultural Statistics Service (NASS) report, feeding co-products to livestock is most common in dairy and cattle feeding operations and less common in hog and cow-calf operations (USDA, NASS, 2007). However, the report also shows that among producers not using co-products, those feeding cattle and hogs are most likely to consider using co-products, presumably if cost-effective opportunities exist.

There are additional co-product characteristics that support inclusion in livestock rations. Cow-calf operators and cattle feeders tend to use co-products when they provide a high energy value for the cost, compared with traditional energy sources like corn or other grains. Dairy and hog operations, on the other hand, value co-products for their protein as well as energy, and use them as cost-effective substitutes for traditional grains and proteins in feed rations.

Nutrients present in corn or other feed grains used to produce ethanol and other starch-based products are concentrated during the milling process by a factor of about 3. For some nutrients, like lysine and methionine, the resulting concentration is a plus, as those amino acids are often at insufficient levels in most common feedstuffs and must be supplemented in livestock rations

(Schingoethe, 2007). Co-products can reduce the need for supplements, potentially reducing the overall feed costs.

However, with other nutrients, the concentration produced during ethanol production can be a negative. Some nutrients, like sulfur, can reach levels toxic to livestock, and means of offsetting the excess nutrient must be utilized. Sulfur can reach excessive levels in wet or dry distillers' grains with solubles when too much sulfuric acid is used in the cleaning process, or if the water used has a naturally high sulfur content (Erickson et al., 2007; Schin-goethe, 2007). High-sulfur water issues can be exacerbated if water on the livestock operation contains sulfur.

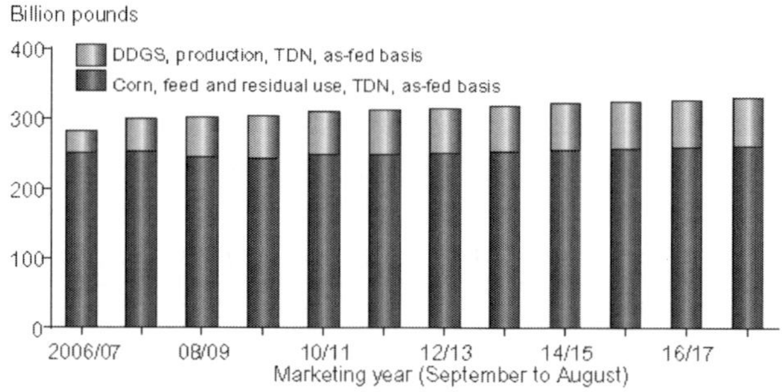

Note: TDN = Total digestible nutrients; DDGS = Dried distiller's grains with solubles.
Source: *USDA Agricultural Projections to 2018*, www.ers.usda.gov/publications/oce091/.

Figure 3. Corn feed use, DDGS production increase feed energy potential (TDNs) .

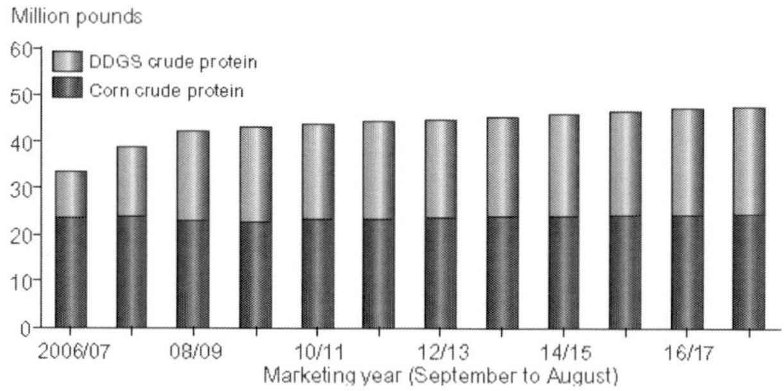

Note: DDGS = Dried distiller's grains with solubles.
Source: *USDA Agricultural Projections to 2018*, www.ers.usda.gov/publications/oce091/.

Figure 4. Crude protein production projections.

For some nutrients, the concentration achieved during milling can be positive or negative, depending on circumstances. Phosphorus, for example, is concentrated as a readily available, highly digestive form, which can reach significant levels in manure. This is a positive if the manure is then spread on land deficient in phosphorus, but can be a negative if it contributes to excessive phosphorus in the soil or contributes to surface water pollution (Schingoethe, 2007).

Ruminants fed rations high in grain content can fall victim to acidosis, a condition in which excessive organic acids are produced in the rumen as a result of fermentation of the starch in grains. Metabolizing these organic acids can damage the liver, which results in liver condemnations and reduces packer revenues from edible offal. Increasing the amount of fiber in rations with roughages can reduce the incidence of acidosis, but at the expense of reduced per-unit energy and protein. However, with little or no loss in energy and a gain in protein content, fiber content in the co-product condensed distiller's solubles is two times greater than the fiber in corn, and corn gluten feed (CGF) has five times the fiber that corn does. Fiber content limits the usefulness of co-products such as CGF in hog and poultry rations. Combinations of co-products in highly intensive cattle feeding rations can reduce the incidence of acidosis (Schingoethe, 2007).

STORAGE, MOISTURE, AND PRICES OF OTHER FEED INGREDIENTS AFFECT ECONOMIC VIABILITY

The price of competing feeds dictates how economical it is to feed co-products. DDG, CGF, and other co-products were used for livestock feed before ethanol production began rapidly increasing in 2003, but the price ratio of co-product to corn was relatively high. This high price ratio was partly due to foreign demand for DDG. Both corn and DDG prices declined 17 percent from 2003 to 2005[3] (USDA, ERS, 2009). During this short period of price declines for DDG, byproducts were viewed as slowing the ethanol production process until they could be removed (Rendleman and Shapouri, 2007).

Growing supplies of co-products and high corn prices opened up a larger market for co-products as livestock feed and led to the current view of their value. Until late in 2008, both corn and DDG prices had increased to record or near-record levels from their respective lows in 2005 and 2004. In 2003, the DDG-to-corn price ratio was well above 1. At times during 2007 and 2008, the ratio declined to levels of about .85, a level at which it is competitive to use DDG instead of corn for feeding to hogs and poultry.

CGF was the most common co-product fed to cattle on both cow-calf operations and feedlots (USDA, NASS, 2007). CGF used in cattle feeding operations had a higher moisture content (52 percent compared to 26 percent) and was less than half the price ($44.22 per ton versus $101.29) compared to CGF used by cow-calf operators. This price difference mostly reflects costs of drying co-products for pasture feeding. Additionally, 36 percent of cattle feeders that fed co-products to their livestock used WDGS. These WDGS have even more moisture than corn gluten feed, ranging from 57 percent (over 40 percent solids) to 60 percent (25-40 percent solids). At $28.28 (25-40 percent solids) to $31.61 (over 40 percent solids) per ton, these were also the cheapest form of co-products (USDA, NASS, 2007).

Some of the differences between these prices and characteristics can be explained by logistics and geographic location. Where ethanol plants and larger cattle feedlots are in close proximity, co-product inclusion in feed rations is more widely practiced. For example, in a Nebraska study, 59 percent of Nebraska cattle feeding operations[4] included co-products in feed rations (Waterbury et al., 2009). At the national level, 36 percent of feedlots fed co-products (USDA, NASS, 2007).

Cattle feeders also use co-products other than distillers' grains and sometimes rely on co-product combinations. In some cases, the combinations can result in higher feeding values for traditionally processed feed ingredients. For example, high-moisture corn was found to have greater value when fed with wet CGF to beef cattle (Erickson et al., 2007). Some combinations being used include CGF with corn, corn oil and roughage, including soybean straw, and Cargill's Sweet Bran.[5]

The cost of drying various co-products is at least partially reflected in various price differentials (figure 5). According to NASS, DDG without solubles were the co-product choice of most dairies (45 percent) and hog producers (44 percent). Feed of this type was characterized by low moisture content—12 and 10 percent, respectively—and was relatively more expensive than other co-products, at average 2006 prices of $118.47 per ton for dairy cattle and $109.49 per ton for hogs (USDA, NASS, 2007). A large proportion of hog producers (37 percent), particularly larger producers, also used DDG with solubles (DDGS). Solubles can add some energy via their extra fat content and they contain a higher share of phosphorus.

In conjunction with prices, the economic viability in distributing and storing co-products will play a large role in determining how extensively they are used. Advancements in logistics and efficiency improvements in the supply chain will be major factors that will affect the demand and price of co-products in the future. Technological improvements that result in more consistent products would lower the cost and increase the demand for these feedstuffs.

International demand for U.S. co-products will also affect domestic supplies and markets. In the 2007 marketing year (September 2007 to August 2008), foreign markets bought about 19 percent of U.S. production. U.S. exports of DDG and DDGS have expanded exponentially to livestock-producing countries, especially countries with relatively high prices for corn and soybean meal. The U.S. Grains Council estimated that the dollar value of U.S. exports of DDG could reach $1 billion in 2008, up from only $76 million in 2002/03, and could reach $3.5 billion for an estimated 14 million metric tons by 2015 (Keefe, 2008). Mexico, expected to import over a million metric tons of U.S. DDG in 2008, is the largest foreign market, followed by Canada, Turkey, Taiwan, South Korea, and Japan (Keefe, 2008).

Improvements in Storage and Transportation Are Keys to Future Use

Co-products have a number of physical characteristics that present logistical issues for livestock producers using them in feed rations. These characteristics, as well as some chemical characteristics, make transporting, handling, feeding, and storing co-products challenging.

The biggest obstacle to providing co-products to livestock feeders is transporting the feeds from the processing plant to the livestock feeding operation. Depending on the co-product's final processed form, varying degrees of difficulty occur in loading, shipping, and unloading. Moisture in the co-product can cause spoiling and depreciate nutritional content, particularly in the summer when temperatures are higher. The shelf life of high-moisture feeds can, therefore, be relatively short.

High moisture content also makes the co-product more difficult to handle for shippers. Shipping a high-moisture feed adds to the total weight of the shipment, and product consistency can cause the co-product to stick to the containers, causing difficulties in the unloading process. Generally, the lower the moisture content, the easier it is to transport co-products.

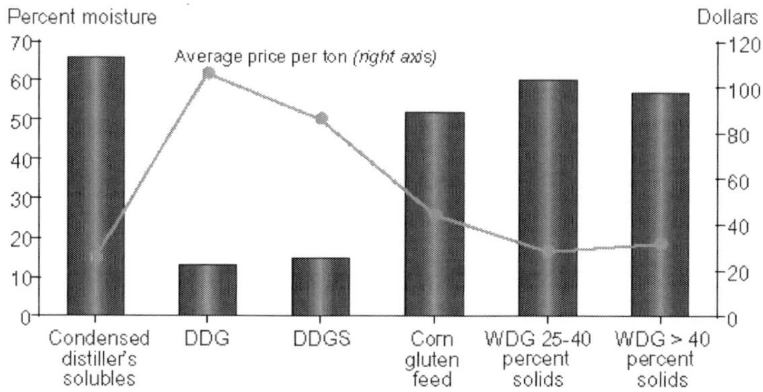

Note: DDG = dried distiller's grains; DDGS = dried distiller's grains with solubles; WDG = wet distiller's grains.

Source: USDA, National Agricultural Statistics Service, http://usda.mannlib.cornell.edu/MannUsda/viewDocumentInfo.do?documentID=1756/.

Figure 5. High-moisture co-products cost cattle feeders less than low-moisture co-products, 2007.

However, removing moisture from co-products requires large, capital-intensive dryers and increases energy costs. In addition to the cost of drying, exposing co-products to high temperatures increases the risk of scorching, which can reduce the nutritional content and the overall value of the feed. In some cases, livestock producers feeding low-moisture co-products find animals exhibit lower feed efficiency, which the producers remedy by putting water back into the feed. Drying costs account for most of the price differential between co-products that differ only in moisture content.

Typically, U.S. livestock operations where co-products are used are larger operations (USDA, NASS, 2007). Such operations can use larger quantities of co-products on a regular basis, providing ethanol plants with a continuous and consistent market for their co-product production. Cattle feeding operations where co-products were used averaged 76 percent larger than all cattle operations (1,276 head vs. 725 head) (USDA, NASS, 2007). Hog operations using co-product feed were over three times as large as average operations (10,957 head vs. 3,256 head). One reason for this size differential is that co-products need to be transported from the ethanol production facility as soon as possible to make room for the next batch of co-products, in order to maintain peak ethanol production. Livestock feeding operations that can make use of full truckloads of co-products are more easily able to work with ethanol plants to obtain co-products on a regular basis at attractive prices or perhaps even contract for delivery.

Based on 2007 NASS survey results, larger cattle feeders near plants may have been the best customer for co-products because (USDA, NASS, 2007):

1. Cattle feeders report prices at a discount compared to dairy, hog, and cow-calf operations.
2. Cattle feeders report using co-products with the highest moisture content.
3. Cattle feeders report that they receive the fewest services from co-product suppliers, compared with other cattle operations, and they are twice as likely to order their feed directly from a processing plant.

Erickson et al. (2007) observed that returns per head decrease with the distance needed to ship a given proportion of Cargill's Sweet Bran used in cattle feeding rations, especially if the distances were at least 100 miles from an ethanol plant. Similar results were observed in another Nebraska study (Waterbury et al., 2009). These findings may also refl ect the ability of cattle to grow to slaughter size more efficiently on feed with WDGS compared with feed efficiency when cattle are fed DDGS, as well as the fact that co-products are less expensive prior to incurring drying costs.

Since the movement of co-products offsite facilitates ethanol production, ethanol manufacturing plants are anxious to market and transport the co-products as quickly as possible. The most economical situation for ethanol producers is to be able to ship large amounts of high-moisture co-products for short distances, thus, avoiding drying costs as well as maintaining nutritional content and limiting spoilage. Livestock operations situated within 100 miles of ethanol plants therefore have an advantage in being able to obtain and use co-products efficiently.

Transportation technologies are constantly being introduced and improved. For example, unit trains (such as 90- or 110-car trains dedicated to a single type of traffic), are being used to ship co-products to specific locations. Some milling/processing plants have built specialized rail track loops and unloading facilities to move their co-products. Some are able to unload special railroad cars without decoupling them, and rotate the entire car upside down, to dump co-products into pits from which they can be redistributed or mixed with other feed ingredients and fed to livestock.

Most U.S. livestock producers buy their co-products on the spot market and prefer not to buy under contracts (USDA, NASS, 2007). Contracting for co-products may result in a more reliable supply source. In a study focusing on Nebraska cattle feeders, who are generally close to ethanol plants, most co-product was sold in 2007 on a fixed-price, annual contract basis (Waterbury et al., 2009). However, in 2008, the absence of contracts between ethanol producers and livestock operations may have been motivated by ethanol plants (Mark, 2009). Cost may not be the only reason producers would contract their co-products; consistency and quality of the co-products are also concerns.

Co-products' quality and content can vary as processors make adjustments in order to optimize production. Because the feed is a byproduct of the processing plant, the nutrient contents of the feed may vary, based on a number of factors in the production process of primary starch related products and oils. However, as the profit margins from ethanol production decline, ethanol producers have made more efforts recently to manufacture feed co-products that are consistent in quality and content, which can sell at a premium.

Education among livestock operators about how to use co-products as animal feed appears to be a determinant in the widespread use of co-products. Although most respondents in a NASS survey stated that lack of availability of co-products was the primary reason why they did not feed co-products to their animals, 5 percent of producers in every livestock category cited lack of knowledge as the primary reason (USDA, NASS, 2007). There is potential for more co-product use if U.S. livestock operators become more knowledgeable about how best to manage ethanol co-products as animal feed.

Considerations for the Future

The use of co-products in feed for U.S. dairy, meat, and poultry production will be affected by a number of competing factors. The development and growth of the ethanol industry will affect the availability of corn for feed, increasing competition in the corn market. As the ethanol industry grows, there will be a corresponding increase in co-products available to substitute for corn in animal feed.

Ethanol production increased 232 percent between 2003 and 2007, rising from 2.8 billion gallons to 6.5 billion gallons (table 2). The Energy Independence and Security Act of 2007 established the Renewable Fuel Standard (RFS) which mandates the use of biofuels at 11.1 billion gallons in 2009 rising to 36 billion gallons in 2022. Some of the 36 billion gallons is to come from cellulosic ethanol and other advanced biofuels. The maximum amount of corn-based ethanol that can be used to meet the RFS increases from 9 billion gallons in 2008, to 15 billion gallons in 2015. If all 15 billion gallons were produced from dry-milling corn, as much as 98 billion pounds of DDG could be produced from 5.6 billion bushels of corn.

Co-products add to the options available to informed livestock producers to develop the best feeding strategies possible. Increased education of and outreach toward livestock and poultry producers about how best to use co-products as animal feed will infl uence how operators adapt the new co-products to future feed rations. New information on co-products' nutritional values, distribution possibilities, and other aspects will be disseminated to the livestock industry through extension offices and publicly available research, among other forms of communication.

REFERENCES

[1] Biomass Research & Development Board. (2008). *Increasing Feedstock Production for Biofuels: Economic Drivers, Environmental Implications and the Role of Research,* December. www.brdisolutions.com/Site%20Docs/Increasing%20Feedstock_revised.pdf/.

[2] Erickson, G. E., Bremer, V. R., Klopfenstein, T. J., Stalker, A. & Rasby, R. (2007). *Utilization of Corn Co-Products in the Beef Industry*, University of Nebraska-Lincoln. August. www.nebraskacorn.org/publications/coproducts_utilization.pdf/.

[3] *Feedstuffs: 2009 Reference Issue and Buyers' Guide*. (2008). Miller Publishing Co., Minnetonka, MN, September 10.

[4] Keefe, D. (2008). *International Trade in DDGS Supply & Demand Outlook 2008-2015*, presented at Workshop on Emerging Issues in Global Meat Trade, USDA, Economic Research Service and Farm Foundation, Washington, DC, September 18-19.

[5] Lawrence, J. (2009). Iowa State University, Ames, IA, Personal communication, January 20.

[6] Mark, D. (2009). University of Nebraska-Lincoln, NE, Personal communication, January 11.

[7] Rendleman, C. M. & Shapouri, H. (2007). *New Technologies in Ethanol Production.* USDA, Office of the Chief Economist, Agricultural Economic Report No. 842, February. www.usda.gov/oce/reports/energy/aer842_ethanol.pdf/.

[8] Renewable Fuels Association. *Historic U.S. Fuel Ethanol Production.* 2005-2009. www.ethanolrfa.org/industry/statistics/#A/.

[9] Schingoethe, D. J. (2007). *Strategies, Benefits, and Challenges of Feeding Ethanol Byproducts to Dairy and Beef Cattle*, presented at the Florida Ruminant Nutrition Symposium, Gainesville, FL. January 30-31.

[10] Service, May. www.ers.usda.gov/Publications/FDS/2007/05May/FDS07D01/.

[11] Sevcik, J. (2006). *Fuel Ethanol and Implications for Animal Agriculture*, presented at the American Meat Institute, October 5. www.meatami.com/ht/a/GetDocumentAction/i/929/.

[12] Shapouri, H. (2009). USDA, Office of Chief Economist, Office of Energy, Personal communication, January 9.

[13] Trostle, R. (2008). *Global Agricultural Supply and Demand: Factors Contributing to the Recent Increase in Food Commodity Prices.* WRS-0801, USDA, Economic Research Service, May.

[14] U.S. Department of Agriculture & Economic Research Service. (2009). Feed Grains Database. Accessed on February 10, at: www.ers.usda.gov/Data/feedgrains/.

[15] U.S. Department of Agriculture & Office of the Chief Economist. (2009). *USDA Agricultural Projections to 2018*, OCE-2009-1, February.

[16] U.S. Department of Agriculture, National Agricultural Statistics Service. *Ethanol Co-Products Used for Livestock Feed*, Sp Sy 6-1(6-07), June 29, 2007.

[17] Waterbury, J. A., Mark, D. R., Thoms, S. M., Erickson, G. E. & Klopfenstein, T. J. (2009). *An Economic Overview of Ethanol Co-product Utilization in Nebraska* Nebraska Beef Cattle Report, University of Nebraska University Extension publication EC858, 2009. http://beef.unl.edu/beefre-ports/200918.shtml/.

[18] Westcott, P. (2007). *Ethanol Expansion in the United States: How Will the Agricultural Sector Adjust?* FDS-07D-01, USDA, Economic Research.

[19] Wisner, R. (2007). Professor Emeritus, Iowa State University, Ames, IA. Personal communication, March 29.

End Notes

[1] Solubles are the liquid portion, or thin stillage, removed from the mash (a mixture of grain, water, and other ingredients) during the process of producing ethanol. Solubles are

essentially a byproduct stream, but, because they do contain some nutrients, the solubles can be included in the distiller's grains co-products.

[2] A ruminant is an animal with a rumen compartment in its stomach. Examples include cattle, sheep, goats, deer, elk, and bison. Unlike animals with a single compartmented stomach (monogastric), ruminants are able to convert cellulose in its various forms into starches and sugars, which can then be metabolized.

[3] Calendar-year basis

[4] These feedlots had an average one-time capacity of 5,760 head and accounting for 91 percent of cattle represented in the survey.

[5] Sweet Bran is a registered trademark of Cargill, Incorporated. It is a dairy feed made from ingredients of the wet corn milling process.

In: Feed Market Dynamics …
Editors A. Stewart and K. Gray
ISBN: 978-1-61942-272-8

Chapter 3

ECONOMIC IMPACTS OF FEED-RELATED REGULATORY RESPONSES TO BOVINE SPONGIFORM ENCEPHALOPATHY

Kenneth H. Mathews

ABSTRACT

Animal and poultry disease outbreaks often lead to new or amended policies and regulations. The economic effects induced by these policies can be much greater and much longer lasting than the immediate effect of the disease outbreak alone. Using Bovine Spongiform Encephalopathy (B SE) as an example, this paper demonstrates the pervasiveness of the effects of restrictive feed policies and regulations, particularly as they relate to meat and bone meal and other protein feeds. Costs evaluated include those assumed by consumers via changes in supplies of secondary and final products; environmental costs associated with disposal of hazardous materials; lost value of products to the rendering industry, including a decline in value of meat and bone meal; and costs of supply disruptions and substitutions within the feed market sector. Benefits from new or amended policies accrue but are not easily measured.

ACKNOWLEDGMENTS

Thanks to the numerous reviewers and others who provided useful comments and technical expertise in publishing this report. Thanks especially to Allen Baker, Mark Ash, and Don Blayney (USDA, Economic Research Service (ERS)); John Fox (Kansas State University); Clark Nardinelli (USHHS, Food and Drug Administration); Jim Schaub (USDA, Office of Risk Assessment and Cost-Benefit Analysis); Kristyn Stone (USDA, Animal and Plant Health Inspection Service); James Wilkus (USDA, Food Safety and Inspection Service); and John Weber and Victor Phillips, Jr., (ERS) for their expertise with technical aspects of the publication process.

POLICY RESPONSES EXTEND DISEASE IMPACTS

The economic effects from animal disease outbreaks include outcomes resulting from policies and regulations designed to mitigate or prevent future disease outbreaks. While the primary motivation for policy responses is to benefit society at large, these policies and regulations can carry significant costs for some sectors that are often pervasive and difficult to identify or quantify, and may have unintended consequences. The distribution of the economic burden of regulations among producers, processors, and consumers varies depending on the producers' and processors' ability to pass on the increased costs and consumers' willingness to pay higher prices for meat. These policy-induced effects can be much greater and much longer lasting than the immediate economic effects of the disease.

Prevention/mitigation policies and regulatory changes related to Bovine Spongiform Encephalopathy (BSE), or mad cow disease, affect a large number and variety of upstream and downstream sectors. Significant among these policies and regulations are those that restrict feeds or feeding, either directly or indirectly, as a means of preventing or mitigating the spread of animal diseases. Using policy and regulatory responses to BSE as an example, this paper demonstrates the range of costs and effects associated with animal disease-related policies and regulations. Costs are easier to estimate than benefits.

The effects from BSE-related feed-policy responses are widespread and extend far beyond the cattle and beef industries. This is due to the number of industries dependent on outputs from the byproduct and rendering industries and feed manufacturing sectors. Affected industries include the cosmetic and pharmaceutical industries (both of which use byproducts, such as gelatin and collagen), feed manufacturing industries, and numerous service and manufacturing industries (which use other animal byproducts, such as enzymes, triglycerides, and other compounds in the manufacture of fatty acids, paints, varnishes, rubber goods, plastics, and lubricants).

Restrictive policies reduce the possible uses of the numerous byproducts and rendered products by cosmetic, pharmaceutical, manufacturing, and other industries. To recoup the lost value, industries must discover new uses for these materials. New regulations can also saddle industries with costs associated with disposal of these hazardous materials in an environmentally benign manner (Informa Economics, 2005; Coffey et al., 2005). In addition, downstream markets for substitutes for the restricted products may be affected, and consumers may be affected via changes in supplies of secondary and final products.

The costs of regulations of byproducts accrue to cosmetic, pharmaceutical, manufacturing, and other industries and their upstream suppliers of intermediate byproducts and offal. For example, costs associated with BSE-related restructuring in upstream (livestock producers, meatpackers, etc.) and downstream (manufacturers of pharmaceuticals, cosmetics, plastics, lubricants, leather goods, etc.) industries are significant. Implications of these costs can vary based on whether the numbers of affected animals are limited (as in the case of BSE) or widespread (as in the case of an outbreak of foot-and-mouth disease). Benefits from animal disease mitigation and prevention policies and regulations, in addition to being difficult to measure, are often spread over a much broader group of beneficiaries, pointing to the need for a broader view of those negatively and positively affected (see, for example, Paarlberg, Lee, and Seitzinger, 2003).

BSE-related policies aimed at preventing the spread of the disease among cattle and its human form, variant Creutzfeldt-Jakob disease (vCJD), to humans have had economic effects on grain and protein feed markets since BSE was first discovered in the United Kingdom (UK) in 1986. This analysis focuses on the potential economic effects that can arise from preventive or mitigative policies restricting or altering the use of animal-based feed ingredients, particularly meat and bone meal (MBM) and other proteins.

BSE-Related Feed Regulations

In the case of BSE, in which outbreaks tend to involve relatively small numbers of livestock, policy and regulatory implementation can still result in large changes in protein feedstuff markets. The changes occur even in countries with no confirmed cases of BSE because virtually every country imposes regulations to prevent the spread of the disease (USHHS, FDA, 1997a and b). Further, in concluding their updated risk assessment of BSE in the United States, Cohen and Gray (2005, p. 35-36)) suggest that "[it] is worth noting that [the risk management measures adopted by the U.S. Department of Agriculture, considered by FDA, or proposed by the International Review Subcommittee] reduce what is already a small exposure [to BSE] in absolute terms." Changes occur in markets for substitutes (e.g., other proteins) and complements (e.g., other feedstuffs fed in conjunction with proteins). Changes in BSE-related feeding policies and regulations also require re-evaluation of the disposal of livestock wastes, as new or expanded uses or means of disposal have to be developed for newly prohibited materials.

A direct impact of implementing preventive and mitigative policies or amended regulations related to animal disease outbreaks can be euthanization of large numbers of livestock. In the short term, a drop in numbers of livestock reduces demand for feeds (grains, proteins, and byproducts) and feeding services (feedlots and other feeding facilities). Disease prevention and mitigation policies and regulations also have a number of indirect effects on feed markets. Restrictions on MBM, for example, can affect prices of other protein feeds.

A Short History of Meat and Bone Meal and BSE

To understand the complex nature of policy-induced effects on feed markets, one should first recognize that animal-based proteins, including MBM and blood meal, are high-quality protein sources rendered from many terrestrial and aquatic animals that are important feeds for livestock worldwide. Prior to 1997, no distinction was made between MBM derived from ruminants (animals whose stomachs have multiple compartments) and that from nonruminants (animals whose stomachs have a single compartment), and MBM represented an additional source of both protein for animal feeds and revenue for renderers and packers. MBM's contribution to total protein supplies had the effect of reducing costs of animal feeds. MBM's contribution to packers/ renderers increased the value of livestock, ultimately increasing cattle prices, while lowering the prices consumers paid for meat. In the United States, as recently as 2000, most MBM was fed to poultry (45 percent), pets (25 percent), swine (15 percent), and cattle (10 percent, nonruminant MBM) (Coffey et al., 2005).

MBM and other animal-derived proteins continue to be important sources of protein for nonruminants in the United States. Vegetable proteins (soybean meal, cottonseed meal, and distillers' byproducts), nonprotein nitrogen (urea), and, increasingly, distillers' grains are often more commonly fed to ruminants because they are generally cheaper sources of protein. These types of feed are especially suitable for ruminants because they can convert vegetable proteins and urea to high-quality protein in the rumen (one compartment of the stomach).

In the United States, most protein in cattle rations is fed to lactating dairy cows and cattle being fattened in feedlots during the final few months prior to slaughter. Dairy cows and calves account for most of the 70 percent of blood meal used by ruminants (Sparks, 2001). Even before FDA prohibited feeding protein derived from ruminants, other than blood products, to ruminants in August 1997, only small amounts of animal proteins were fed to ruminants in the United States, primarily to dairy cattle, because of the relatively high cost of this type of feed. Range cattle, which are on pasture most of their lives, typically receive little protein supplement of any kind.

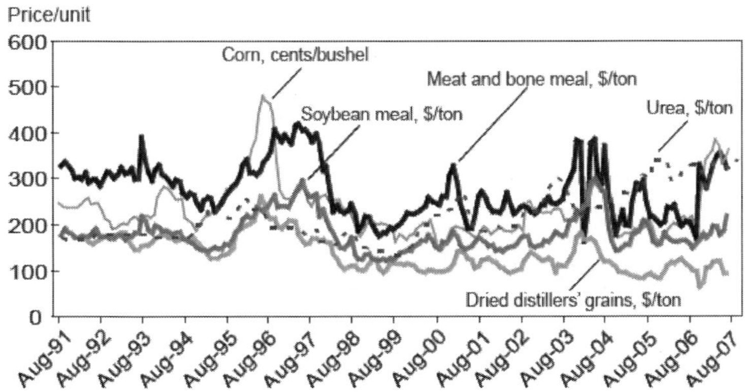

Source: Compiled by USDA, Economic Research Service, with data from USDA, Agricultural Marketing Service, and U.S. Department of Commerce, Bureau of Census.

Figure 1. Selected monthly U.S. feedstuff prices.

In the European Union (EU), inadequate production of protein feeds and tariff structures on grains were such that animal proteins were priced more competitively with vegetable proteins (Hasha, 2002). EU producers made wide use of MBM in cattle feed rations before the EU feed ban. This practice, along with changes in the rendering process in the EU, is thought to have contributed to the spread of BSE, particularly in the UK. In an ex ante analysis, it was estimated that the 2001 EU ban against MBM in any animal feed would cause the EU to import an additional 1.5 million more tons of soymeal per year to replace meat and bone meal in livestock feed rations (USDA, OCE, 2002).

Even before the discovery of BSE in North America, some U.S. poultry producers had reduced their use of MBM in poultry feed rations, in part, as a response to BSE-related problems in the EU. U.S. producers were also likely heavily influenced by the 1996 UK announcement of a possible connection between BSE and vCJD in humans. As MBM use in the United States declined, prices for substitute sources of high-quality livestock protein feed-stuffs increased, although prices for proteins continued to move together, despite regulatory effects on MBM (figures 1, 2).

Benchmark BSE Events

Four periods have been marked by benchmark BSE events that resulted in regulatory/policy responses (Mathews et al., 2003; Mathews et al., 2006). In 1986, cattle were found to be infected with a new Transmissible Spongiform Encephalopathy (TSE), a group of neurological diseases affecting several mammalian species. First discovered in the UK, the disease—BSE—was considered by most countries to be an animal health issue, and various restrictions, including those on trade, were imposed on the basis of protecting animal health.

In 1996, BSE-related regulations shifted significantly with the discovery of B SE-infected cattle in EU countries other than the UK (France, Ireland, and Portugal) and, more importantly, with the announcement in March 1996 of a potential link between vCJD in humans and BSE in cattle. In 2000, native-born BSE cases were discovered in other EU countries[1] and in countries outside the EU (e.g., Japan in 2001). Lastly, in 2003, BSE was discovered in a cow in Canada (May) and in a cow in the United States that had been imported from Canada (December).

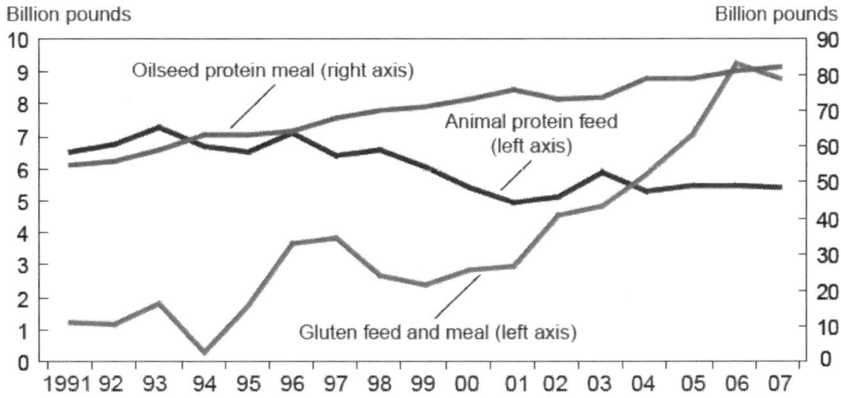

Source: Compiled by USDA, Economic Research Service, with data from USDA, Agricultural Marketing Service, and U.S. Department of Commerce, Bureau of Census.

Figure 2. Annual consumption of selected U.S. feedstuffs.

The sharp MBM price decline in 1997 in the United States (see figure 1) was due to the 1997 feed ban, which followed the 1996 UK announcement of a BSE connection with vCJD in humans. Effects of the spread of BSE to other countries are reflected in the increased rate of decline in U.S. MBM production in 2000. The 2003-04 round of BSE-related feed regulations[2] resulted in another sharp decline in U.S. MBM prices in 2004 and 2005.

In the United States, BSE-related feed policies began with an emergency ban on importation of all ruminants and MBM from the UK in 1989 (Coffey et al., 2005). An extension of the ban followed on December 6, 1991, restricting all high-risk products from countries with known cases of BSE. U.S. MBM prices continued a decline that began at the end of the drought in 1988 and reached a nominal low in spring 1995 with a per ton amount that ranged from the upper $1 50s to the lower $1 60s. While there is no evidence that regulations seriously affected domestic MBM markets, the early part of the decline in U.S. MBM prices may have been induced in part by BSE-related market activity.

The 1996 announcement of a potential link between vCJD in humans and BSE in cattle elevated BSE-related concerns and shifted them from an animal health issue to a human health issue. This shift also resulted in significant feed regulations being imposed, particularly aimed at MBM. In August 1997, the U.S. Food and Drug Administration (FDA) implemented a regulation prohibiting the use of mammalian protein, primarily MBM, in the manufacture of ruminant feeds (USHHS, FDA, 1 997a and b).[3] These mammalian protein sources were excluded from feed because the suspected infective abnormal prions are thought to be transmitted in MBM rendered from infected cattle. Prior to the August 1997 ban, MBM could be used in place of or along with soybean meal, cottonseed meal, and other proteins as a source of protein in ruminant rations.

U.S. MBM prices declined by half over the next 2 years, from a high of $431 per ton in April 1997, attained in part due to drought conditions in 1996 that drove grain prices to record highs, to a low of $187 per ton in March 1999. Another significant development during this period was segmentation of the MBM market into two markets beginning in summer 1997, porcine- (derived primarily from hogs, horses, and other nonruminant animals) and ruminant- derived MBM (figure 3). Porcine MBM averaged a premium of $15.78 per ton over the period January 1998-December 2003 (Coffey et al., 2005), just before discovery of the first U.S. case of BSE.

In assessing the economic effects of the 1997 ban of MBM in ruminant feed, FDA estimated that U.S. MBM prices would decline by between $25 and $100 per ton (USHHS, FDA, 1997a) and accepted Sparks Companies, Inc.'s estimate of a decline of $68.27 per ton (USHHS, FDA, 1997b, p. 30970). FDA rejected the assumption that other protein feeds, like soybean meal, would substitute for all MBM used in ruminant feeding, and instead assumed that only 10-15 percent would be replaced, at an estimated cost of $8 million (USHHS, FDA, 1997b, pp. 3097 1-30972). Annualized private-sector costs of measures implemented due to the 1997 feed ban regulations were estimated at between $21.4 million and $48.2 million (USHHS, FDA, 1 997a). These estimates pertained almost exclusively to the rendering industry and markets for byproducts and rendered products. Revised FDA estimates indicated that the ban would ultimately cost the U.S. private sector approximately $53 million per year, reflecting $44 million in direct compliance costs to beef, rendering, and by product industry sectors (including annualized capital and operating costs), $171 million in lost value of products to the rendering industry (primarily a decline in value of MBM), and a gain of $163 million, through lower feed costs, to producers of nonruminant animals (USHHS, FDA, 1997b).

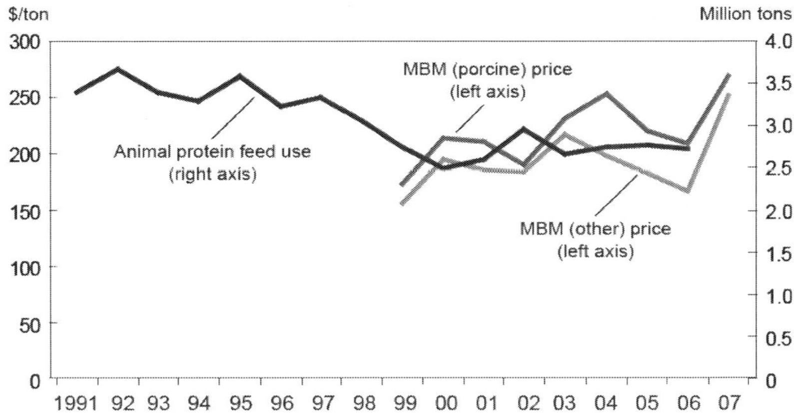

Source: Compiled by USDA, Economic Research Service, with data from USDA, Agricultural Marketing Service, U.S. Department of Commerce, Bureau of Census, and National Renderers Association.

Figure 3. U.S. meat and bone meal (MBM) prices and animal protein feed use.

In December 2000, the United States banned imports of all rendered products from Europe for feeding, regardless of species of origin (Coffey et al., 2005). The United States also increased its surveillance by increasing BSE sampling and broadening sampling procedures to account for any regional differences in BSE incidence. These measures were characterized by increasing MBM prices and, in 2002, the narrowest price spread between porcine and ruminant MBM since reporting of the two price series began. MBM prices had recovered somewhat by 2003, with prices reaching a high of $395 in November 2003 (see figure 1).

The United States announced a series of new regulations shortly after the December 2003 BSE discovery. These additional regulations were designed to enhance protections against the spread of BSE and to reassure consumers that beef was safe to eat. Enhanced testing objectives were also implemented in the United States to determine the extent to which BSE was present in the U.S. cattle population.

In late December 2003 and January 2004, both the FDA, the agency in charge of animal feed contents, and USDA's Food Safety and Inspection Service (FSIS), the agency responsible for animal welfare and slaughter, imposed additional regulations to prevent the spread of BSE to humans and animals (USDA, FSIS, 2004a). These interim proposed rules, four from each agency, were followed by an FDA rule announced in October 2005 that proposed to prohibit the use of certain cattle origin materials in the food or feed of all animals (USHHS, FDA, 2005b).

In January 2004, MBM prices dropped to $173 in response to the U.S. BSE confirmation. But by April 2004, MBM prices had recovered to a peak of $397 per ton. The regulations FDA posted during summer 2004 sent MBM prices into another decline, reaching a low of $187 per ton in October 2004, a level not seen since.

Other Economic Effects

A number of feeds serve as sources for protein in livestock rations. In addition to oilseed and animal proteins, alfalfa hay and other alfalfa products are sources of high-quality protein, especially for ruminants. Wet and dried distillers' grains also can be used as protein feeds, again, especially for ruminants. Higher prices for protein feeds would result in some limited substitution of feed grains for some protein, especially for hogs and poultry, and alfalfa for some protein, especially for cows. However, substitution of corn for protein could be limited by the use of corn to produce ethanol, which provides support for feed grain prices.

Another factor in the corn-protein substitution scenario is that a bushel of corn (56 pounds) yields about 2.7 to 2.8 gallons of ethanol and only 17.4 pounds of distillers' grains. Thus, in terms of total feed usage, ethanol production reduces the supply of energy feeds by two-thirds (the 56:17.4 conversion). Supplies and use of wet and dried distillers' grains, byproducts of ethanol production, are rapidly increasing as ethanol production increases. The increased use of these increased supplies of distillers' grains would likely have offsetting effects on any increases in the prices of other protein feeds stemming from regulatory restrictions on MBM use. Because the nutrient profile of distillers' grains exhibits high levels of both protein and energy, the increasing use of distiller's grains at attractive prices is a confounding factor in any BSE-related, and, therefore, protein-related analysis.

The regulatory measures announced by both FDA and FSIS in late December 2003 and January 2004 required slaughterhouses to remove additional products from cattle and implicitly challenged industries to find new uses for or means to dispose of materials on the expanded list of cattle materials prohibited from animal feeds (CMPAFs, formerly SRMs, or specified risk materials) and other materials banned from human or animal consumption. As a result, production costs increased and revenues declined for producers and processors of beef products and byproducts.

If future regulatory activities were to implement enhancements originally proposed by FSIS and FDA in 2005 or further enhancements, other protein meal markets could be significantly affected. For example, based on earlier analyses, if the extreme version of the MBM ban were adopted (no mammalian protein fed to any animals), estimates of price increases for other protein sources range as high as 100 percent (Coffey et al., 2005). Many view such an increase as unlikely, given the current MBM share of total protein feeds and sources, which most recently, includes increasing quantities of distillers' grains. Preliminary research at ERS (Mathews and McConnell, 2008) suggests more modest price changes for other proteins for the 13.6-percent supply shock assumed by Coffey et al. (2005).

In addition, Coffey et al.'s 13.6-percent decrease in protein meal supplies was based on MBM shares of total proteins prior to the 1997 feed ban. More recently, MBM represents a less-than-8-percent share of total proteins (excluding proteins in such feedstuffs as alfalfa hay, distillers' grains, and corn gluten feed). Given a less-than-8-percent share, a complete ban would likely result in a much smaller increase in soybean meal and other protein prices than Coffey et al.'s estimate of 100 percent.

Coffey et al. estimated that per animal costs of feeding a beef steer would increase from $18.51 (includes $4.50 for increased feed costs from substitution effects for a ruminant protein to any farmed animal) to $23.61 (includes $9.60 for increased feed costs from substitution effects for a ban against feeding any animal protein to any farmed animal) (table 1). To the extent that these estimates are based on the inflated impact of MBM on soybean meal prices and an extreme scenario, estimates of the impact of FDA's (2008) Final Rule on steer feeding costs are expected to be much lower.

Until FDA published its 2008 Final Rule, proteins manufactured from ruminants and other animals could be fed to hogs and poultry because neither species is known to be affected by any known TSE. Because current and additional, but related, interim rules imposed by FSIS and FDA would change current feed manufacturing practices in significant ways (USDA, FSIS, 2004a and b; USHHS, FDA, 2005a and b; USHHS, FDA, 2008), the FDA Final Rule went through an extended comment and analysis period. In the meantime, the potential economic effects of these regulatory actions on the feed manufacturing industries were estimated, including those by FDA (USHHS, FDA, 2005b, 2008), Coffey et al. (2005), and Informa Economics (2005). However, analyses for the most extreme scenario became moot with the publication of FDA's Final Rule on April 25, 2008, a modified version of the October 2005 proposed rule (see footnote 1), which bans feeding cattle materials prohibited in animal feeds (CMPAF) in feeds for all animals. When this Final Rule is implemented in 2009, feed substitution impacts will likely be relatively small in the aggregate (Mathews and McConnell, 2008).

However, in addition to increased feed costs and other downstream costs, increased restrictions on feed ingredients have introduced an additional round of disposal issues and costs. For example, as FDA noted, renderers, who once picked up dead animals for free, become less inclined to pick up animals that cannot be rendered or will pick them up only at costs estimated at $100 or more per head (USHHS, FDA, 2008; Sparks, 2002). Disposal of additional SRMs/CMPAFs becomes an issue for packers, renderers, incinerators, and landfills in terms of increased costs and environmental concerns. Additional disposal costs for a 1,000-pound animal from which SRMs/CMPAFs can no longer be removed are estimated at $18.75 per head (Coffey et al., 2005). Estimates of increased costs of expanded SRM/CMPAF regulations, which include costs of disposing of SRMs/CMPAFs, range from $2.16 per head (cattle under 30 months old) to $6.77 per head for older animals (see table 1).

Table 1. Economic effects of BSE-related feed regulations banning meat and bone meal proposed by the U.S. Food and Drug Administration on selected livestock-related sectors

Affected sector	Per unit effects	Estimated aggregate annual effects
	Dollars	*Million dollars*
Increase in meat and bone meal prices[a]	25-100/ton	8
Increase in definition of specified risk material[b]	2.16-6.77/head	195
Increase in costs of feeding cattle[b]	23.61/head	637
Net of loss to rendering and gain to nonruminant livestock feeding[a]		53
Export losses[b]		3,200-4,700

Source: Compiled by USDA's Economic Research Service from [a]USHHS, FDA (1997b) and [b]Coffey et al. (2005).

CONCLUSION

Meatpackers and renderers are among the first downstream users of cattle and beef products and byproducts affected during a major animal disease outbreak as regulations prohibit the entry of the meat of diseased animals into the food system. This results in supplies of raw materials (animals or meat) that cannot be disposed of through normal marketing channels. Policies and regulations implemented in the aftermath of a disease outbreak are often extensive, affect related industries, and have economic effects that last much longer than those of the disease outbreak.

While a number of costs associated with regulatory responses have been estimated, particularly for the rendering and other cattle and beef industry sectors, benefits of regulation have been neglected in this research. The most obvious benefit is reduction of risk from animal disease—in this case, risk to animals from BSE and risk to humans of vCJD. However, the incremental value of this risk reduction has not been estimated. Few would argue that reducing the risks from either the spread of BSE among U.S. cattle or to humans in the form of vCJD is not worth pursuing. However, some in the industry and other analysts have begun to ask about the tradeoffs between marginal reductions in these risks and the costs incurred in achieving these risk reductions. While benefits from risk reduction, trade gains, some gains to producers not directly affected by disease losses, and some consumer gains have been estimated, others have yet to be appropriately identified, much less valued.

REFERENCES

[1] Coffey, B., Mintert, J., Fox, S., Schroeder, T. & Valentin, L. (2005). *The Economic Impact of BSE on the U.S. Beef Industry: Product Value Losses, Regulatory Costs, & Consumer Reactions*, MF-2678, Kansas State University Agricultural Experiment Station and Cooperative Service.

[2] Cohen, J. T. & Gray, G. M. (2005). *Harvard Risk Assessment of Bovine Spongiform Encephalopathy Update: Phase IA*, Harvard Center for Risk Analysis, Boston, MA, October 31.

[3] Hasha, G. (2002). *Livestock Feeding and Feed Imports in the European Union—A Decade of Change*, Electronic Outlook Report No. FDS-0602-

01, U.S. Department of Agriculture, Economic Research Service. www.ers.usda. gov/publications/fds/july02/fds0602-01/, accessed on September 28, 2006.

[4] Informa Economics. (2005). *Economic Impacts of Proposed Changes to Livestock Feed Regulations*, McLean, VA.

[5] Mathews, K. H., Jr., Bernstein, J. & Buzby, J. C. (2003). "International Trade in Meat/Poultry Products and Food Safety Issues," chapter 4 in *International Trade and Food Safety: Economic Theory and Case Studies*, J.C. Buzby (ed.), Agricultural Economic Report No. 828, November. www.ers.usda.gov/publications/aer828/, accessed on July 31, 2008.

[6] Mathews, K. H., Jr., & McConnell, M. (2008). "*Protein Feed Market Effects of Regulatory Responses to Bovine Spongiform Encephalopathy*," symposium presentation at the Joint Northeast Agricultural and Resource Economics Association/Canadian Agricultural Economics Association meetings in Quebec, Canada, June 29-July 2.

[7] Mathews, K. H., Jr., Vandeveer, M. & Gustafson, R. A. (2006). *An Economic Chronology of Bovine Spongiform Encephalopathy in North America*, LDP-M-143-01, U.S. Department of Agriculture, Economic Research Service, June. www.ers.usda.gov/publications/ldp/2006/06jun/ ldpm14301/ldpm14301.pdf, accessed on July 31, 2008.

[8] National Renderers Association. *Render*, various issues, Sierra Publishing, Camino, CA.

[9] Paarlberg, P. L., Lee, J. G. & Seitzinger, A. H. (2003). "Measuring Welfare Effects of an FMD Outbreak in the United States," *Journal of Agricultural and Applied Economics 35*(1), 53-65.

[10] Sparks Companies, Inc. (Sparks). (2002). *Livestock Mortalities: Methods of Disposal and Their Potential Costs*, McLean, VA.

[11] Sparks Companies, Inc. (Sparks). (2001). *The Rendering Industry: Economic Impact of Future Feeding Regulations*, McLean, VA.

[12] U.S. Department of Agriculture, Economic Research Service. (2001). "Dissecting the Challenges of Mad Cow and Foot-and-Mouth Disease," *Agricultural Outlook*, AGO-283, August. www.ers.usda.gov/ publications/agoutlook/aug2001/ao283c.pdf, accessed on July 31, 2008.

[13] U.S. Department of Agriculture, Food Safety and Inspection Service. (2004a). "*USDA Issues New Regulations To Address BSE*," January 8. http://www. fsis.usda.gov/oa/news/2004/bseregs.htm, accessed on September 26, 2006.

[14] U.S. Department of Agriculture, Food Safety and Inspection Service. (2004b). "Prohibition of the Use of Specified Risk Materials for Human Food and Requirements for the Disposition of Non-Ambulatory Disabled Cattle; Meat Produced by Advanced Meat/Bone Separation Machinery and Meat Recovery (AMR) Systems; Prohibition of the Use of Certain Stunning Devices Used To Immobilize Cattle During Slaughter; Bovine Spongiform Encephalopathy Surveillance Program; Interim Final Rules and Notice," *Federal Register 69*(7), 1861-74, January 12.

[15] U.S. Department of Agriculture, Office of the Chief Economist (USDA, OCE). (2002). *USDA Agricultural Baseline Projections to 2011*, Staff Report WAOB-2002- 1, prepared by the Interagency Agricultural Projections Committee. www.ers.usda.gov/publications/waob021/, accessed on August 11, 2008.

[16] U.S. Department of Health and Human Services, Food and Drug Administration (USHHS, FDA). (1997a). "Substances Prohibited From Use in Animal Food or Feed; Animal Proteins Prohibited in Ruminant Feed; Proposed Rule," *Federal Register 62*(2), 552-83, January 3.

[17] U.S. Department of Health and Human Services, Food and Drug Administration (USHHS, FDA). (1997b). "Substances Prohibited From Use in Animal Food or Feed; Animal Proteins Prohibited in Ruminant Feed," *Federal Register 62*(108), 30936-30978, June 5.

[18] U.S. Department of Health and Human Services, Food and Drug Administration (USHHS, FDA). (2005a). "Use of Materials Derived From Cattle in Human Food and Cosmetics," *Federal Register 70*(172), 53063-53069, September 7, www.fda.gov/OHRMS/DOCKETS/98fr/05-17693.htm, accessed on September 26, 2006.

[19] U.S. Department of Health and Human Services, Food and Drug Administration (USHHS, FDA). (2005b). "Substances Prohibited From Use in Animal Food or Feed; Proposed Rule," *Federal Register 70*(193), 58570-58601, October 6.

[20] U.S. Department of Health and Human Services, Food and Drug Administration (USHHS, FDA). (2008). "Substances Prohibited From Use in Animal Food or Feed; Final Rule," *Federal Register 73*(81), 22720-22758, April 25, 2008. http://edocket.access.gpo.gov/2008/pdf/08-1180.pdf, accessedon July 10.

End Notes

[1] In 1996, Portugal reported 31 BSE cases, France reported 12 cases, and Ireland reported 74 cases of BSE. By 2000, Portugal reported 150 cases, France 162 cases, and Ireland 149 cases.

[2] USDA, Food Safety and Inspection Service (FSIS) announced four interim final rules on December 30, 2003: (1) changed sampling and testing procedures, (2) expanded the list of specified risk materials, (3) further limited material that could be labeled as "meat" from Advanced Meat Recovery (AMR) procedures, and (4) banned air-injection stunning (http://www.fsis.usda.gov/ news _& _events/nr _0 10804 _0 1/index. asp). FDA went further in closely related Interim Final Rules that banned from FDA-regulated human food, (including dietary supplements) and cosmetics: (1) any material from "downer" cattle, animals that cannot walk, (2) any material from "dead" cattle, those that die before reaching the slaughter plant, (3) Specified Risk Materials (SRMs) that are known to harbor the highest concentrations of the infectious agent for BSE, such as the brain, skull, eyes, and spinal cord of cattle 30 months or older, and a portion of the small intestine and tonsils from all cattle, regardless of their age or health, (4) mechanically separated beef, which may contain SRMs, while allowing meat obtained by AMR, an automated system for cutting meat from bones, because USDA regulations do not allow it to contain SRMs. The FDA human food interim final rule was originally published in July 2004 (http://www.fda.gov/ OHRMS/DOCKETS/98fr/04-15882. pdf) and later amended in September 2005 (www.fda.gov/OHRMS/ DOCKETS/98fr/05-17693.htm). Additional measures were announced later (October 2005) that proposed to prohibit the use of certain cattle origin materials in the food or feed of all animals (www.fda.gov/OHRMS/ DOCKETS/98fr/05-20196.pdf). FDA's Final Rule was published April 25, 2008 (http://edocket.access.gpo. gov/2008/pdf/08-1 180.pdf). This rule further expands the list of banned materials. Specifically, the rule prohibits the use of certain cattle origin materials in the food or feed of all animals: (1) the entire carcass of BSE-positive cattle, (2) brains and spinal cords from cattle 30 months of age or older, (3) the entire carcass of cattle not inspected and passed for human consumption that are 30 months of age or older from which brains and spinal cords were not removed, (4) tallow derived from B SE-positive cattle or other materials prohibited by the rule, and (6) mechanically separated beef that is derived from the materials prohibited by the rule.

[3] Mammalian proteins excluded from this prohibition included "inspected and processed meat products that have been cooked and offered for human consumption and further heat processed for feed (such as plate waste and used cellulosic food casings), milk products, and products whose only mammalian protein consists entirely of porcine or equine protein" (USHHS, FDA, 1997b).

In: Feed Market Dynamics …
Editors A. Stewart and K. Gray

ISBN: 978-1-61942-272-8

Chapter 4

LIVESTOCK FEED COSTS: CONCERNS AND OPTIONS

Geoffrey S. Becker

SUMMARY

Livestock producers in 2008 have seen sharply higher feed costs, fueled by competing use demands for corn and soybeans and by higher energy prices. Some analysts argue that current public policies, including financial incentives that divert corn from feed uses into ethanol production, exacerbated if not caused these higher costs. Other factors, which some believe to be at least as significant, include crop production declines due to weather, and higher global demand for commodities. Proposed options aimed at easing the impacts of higher feed costs include changes in ethanol incentives, use of conservation land for forage use, and direct aid to producers.

ECONOMIC SITUATION[1]

Production costs through the first half of 2008 climbed in every segment of animal agriculture. The main driver was feed, which may account for 60%-70% of total livestock production costs in any given year. Overall, total U.S. feed expenses were forecast to reach a record-high $48 billion in 2008, a jump of nearly $10 billion or 26% over 2007 — a year that was $6.7 billion higher than 2006. Corn accounts for about 90% of feed grains used for feed, and soybean meal is the principal oil crop product used as feed, according to the U.S. Department of Agriculture's (USDA's) Economic Research Service (ERS). Farm-level corn prices have jumped, from an average of $3.04 per bushel (bu.) in the 2006/2007 marketing year (September 1-August 31) to a record $4.20 per bu. in the 2007/2008 marketing year and a projected $5.00-$6.00 in 2008/2009 — a boon for farmers who grow the crops, but bane for animal producers who must buy them.

Soybean farm prices averaged $6.43 per bu. in 2006/2007 and climbed to an estimated record of $10.15 per bu. in the 2007/200 8 marketing year. USDA projects that average farm prices for soybeans could reach $11.60 to $13.10 per bu. in 2008/2009. That has pushed up average soybean meal prices from $205 per short ton in 2006/2007 to $335 for 2007/2008 and a projected $330 to $390 per ton in 2008/2009.

For many animal producers, returns from their own sales were not covering these higher production costs. Total feed costs for finishing a feeder pig in Iowa were about $42, or 30% of the total cost of production, in May 2006; they reached $70 and 46% of total production costs in May 2008. From November 2007 through April 2008, the average loss for each Iowa feeder pig finished for slaughter were nearly $22 per head.[2]

Cattle feeders (who feed young cattle up to slaughter weight) in the Southern Plains, where the major U.S. cattle feeding states include Texas, Nebraska and Kansas, have been in the red since June 2007, according to the Livestock Marketing Information Center (LMIC). Southern Plains feeders have been losing, on average, over $100 per head for much of 2008.[3] CattleFax, which provides market analysis for the cattle industry, estimates that the total cost to "produce cattle and beef from pasture to plate" increased from $726 per head in 2005 to $1,131 in 2008, or by 56%.[4]

Dairy producers have experienced cost pressures, although farm milk prices have remained relatively firm in 2008 due to strong, export-fueled demand. Milk industry analysts also noted that feed costs have moderated somewhat from earlier levels, and that cost impacts may have been muted because "[a] large share of dairy feed in 2008 seems to have either been grown on the dairy or contracted in advance."[5]

A short-term effect when grain prices spike can be more production of red meat and poultry, as herd sizes are reduced and/or as more animals and milk are sold to maintain cash flow to cover higher prices. This can depress farm (and wholesale) prices at least temporarily, further exacerbating the cost-price squeeze. If current market conditions persist, meat supplies will decline — and prices rise — through producer attrition and reduced capacity. Retail prices, too, will eventually adjust accordingly.[6]

"To date, rising costs have largely been absorbed by livestock and poultry producers, often with significant financial loss. However, higher costs of production will ultimately have to be reflected in higher prices for meat, milk, and eggs at retail counters in the United States and elsewhere. This adjustment process is complex, lengthy, painful, and not without unintended consequences," Lawrence and others wrote in *Choices* (see footnote 1). They added that "speed of adjustment will vary significantly as industries with shorter production cycles, such as poultry, are able to respond in a matter of months whereas adjustments in industries with longer production cycles, such as beef, can take a period of several years."

Reasons for Higher Feed Costs[7]

Strong Economic Growth in Developing Countries

Rising commodity prices, including for feed, are rooted in a confluence of factors, some with short-term impacts, but others longer-term. An example of the latter is steadily increasing world population, boosted by robust growth in purchasing power, especially in developing countries such as China and India. Analysts believe this trend could mean a sustained increase in global demand for more and different kinds of food — including, as household incomes improve, higher-value foods like meat and dairy products. This dietary shift in turn implies increased demand for animal feeds.

Weather-Related Crop Shortfalls

Global grain production declined in 2005 and 2006, cutting into existing stocks and reducing exportable supplies. Crops here as well as in Australia, Canada, the European Union, Eastern Europe, and some countries of the former Soviet Union were reduced mainly due to poor weather conditions. On the other hand, weather may have had less of an impact on corn than on food crops like wheat and rice: world corn output was expected to increase from 712 million metric tons (MMT) in 2006/2007 to 790 MMT in each of 2007/2008 and 2008/2009. In the United States, the world's leading producer, 2008 production was estimated by USDA at 12.1 billion bushels, approximately 1 billion bushels below the previous year. Among the reasons cited were spring rains and flooding followed by a dry August in the Corn Belt.[8]

Surging U.S. Exports

Since January 2002, the U.S. dollar has declined in value relative to the currency of U.S. export competitors (e.g., Canada, Australia, or the EU) and grain importing nations (e.g., Japan, Taiwan, etc.), effectively making U.S. grain exports cheaper and, therefore, more competitive, despite the rise in per-bushel prices. In terms of year-to-year export volumes, U.S. corn exports are projected up nearly 18% to a record of almost 2.5 billion bushels in the 2007/2008 marketing year.

Other Contributing Factors

Former USDA Chief Economist Keith Collins cited a number of other factors that have contributed to feed (and other food) price increases. These include higher demand for feed caused by increasing numbers of fed animals in the United States; a two-year rise in corn production costs from $186 per acre in 2005 to nearly $230 in 2007 due mainly to higher energy-related (fertilizer and fuel) costs; and actions by a number of foreign governments to insulate their own markets from high commodity prices, such as limiting exports of their own agricultural commodities and subsidizing domestic production and/or consumption.[9]

Government Biofuels Policy

Perhaps the most widely-debated factor has been government biofuels policy. Current U.S. biofuel production is almost entirely corn-based ethanol — nearly 6.5 billion gallons of corn-ethanol were produced in 2007, compared with an estimated 450 million gallons of primarily soybean-based biodiesel. The Energy Independence and Security Act of 2007 (EISA; P.L. 110-140) extended and substantially expanded the existing Renewable Fuel Standard (RFS), a usage requirement mandating that an increasing volume of biofuels be blended with conventional fuels. The RFS mandates the use of at least 9 billion gallons of biofuel in U.S. fuel supplies in 2009, more than doubling to 20.5 billion gallons by 2015 and to 36 billion gallons by 2022. The U.S. biofuels sector is also supported by a tax credit of 51 cents for every gallon of corn- based ethanol blended in the U.S. fuel supply (to drop to 45 cents in 2009), and an import tariff of 54 cents per gallon of imported ethanol. In addition, several federally-subsidized grant and loan programs assist biofuels research and infrastructure development.

Of the 160 U.S. ethanol plants — with total capacity now in excess of 9 billion gallons per year — only about a dozen use feedstocks other than corn (or milo or barley, also feed grains).[10] USDA estimates that in crop year 2007/2008, about 23% of the U.S. corn crop will have been used to produce ethanol; this share is projected to grow to more than 33% in 2008/2009. Although high corn prices have slowed plans for expanded capacity, plants already on line are expected to operate so long as they can cover their variable costs. The runup in corn demand, which is expected to be long-lasting, has directly sparked substantially higher corn prices to bid available supplies away from other uses like livestock feed, according to a number of agricultural economists.

Corn ethanol byproducts — now mainly distiller's grains from dry-mill ethanol plants but also corn gluten feed and corn gluten meal — can be fed by livestock producers to help offset higher feed corn prices. According to the National Corn Growers Association, the equivalent of an additional 1 billion bushels of feed will be available in the coming year through ethanol byproducts.[11] Because of the cost and difficulty of drying and shipping, distiller's grains are best utilized by feeders closest to ethanol plants. Research indicates that distiller's grains should constitute no more than 3 5-40% of feedlot cattle rations, 10%-20% of dairy cow rations, and 10% for hogs, broiler chickens, and turkeys, according to Anderson.

Reviewing several different studies and economic models, Collins concluded that implied changes in the price of corn due to its use in ethanol might range from 25% to 60%. A 2008 study by the Food and Agricultural Policy Research Institute (FAPRI) attempts to measure the pure and joint price effects of the U.S. biofuels RFS and the tax credits, suggesting that joint implementation of both the RFS and the tax credit supports corn prices by about 20%. A substantial portion of corn price effects are likely transmitted to the soybean market via competition for land, primarily in the Corn Belt where soybeans and corn are both widely grown. A study by the Center for Agricultural Research and Development (CARD) found that, jointly, the RFS and tax credit supported corn prices by 16%. FAPRI and CARD found the impacts of the tax incentives to be highly dependent on petroleum prices; higher petroleum prices substitute for government incentives and diminish their relative impact on corn prices. Neither study included the effects of grants and subsidized loans available for biofuels research and infrastructure.

OUTLOOK AND OPTIONS

The U. S. response to recent developments in grain markets is of keen interest to U.S. meat, dairy and poultry producers, who have long been wary of any government policies that can raise feed prices. Congressional committees in 2008 have held a number of hearings on the causes and impacts of rising commodity costs. The House Agriculture Committee convened six hearings between mid-May and mid-September 2008 on commodity futures trading and related matters, including proposals to increase oversight by the Commodity Futures Trading Commission of futures markets, where, some critics argue, excessive speculation added significantly to the 2008 price spikes in oil and other commodities; others counter that futures traders simply react to market conditions. Economists seem to agree that U.S. government initiatives are unlikely to have much impact on such price drivers as foreign population and income growth and concurrent demand for commodities, weather-related production shortfalls, and inflation generally, at least in the short term. However, a number of options have been offered to address other perceived reasons for high feed prices.

Reduce Domestic Ethanol Incentives

Several bills propose to reduce ethanol production incentives by repealing, lowering, or freezing the RFS, the biofuels tax credit, and/or the ethanol import tariff, such as H.R. 5911, H.R. 5986, H.R. 6137, H.R. 6183, H.R. 6324, S. 3031, and S. 3080. Outside of Congress, attention focused on an April 25, 2008 letter to the U.S. Environmental Protection Agency (EPA) from Texas Governor Rick Perry, requesting a waiver of 50% of the 2008 RFS for grain-derived ethanol, now at 9 billion gallons. The letter cited EPA's authority, under Sec. 211(o) of the Clean Air Act, to approve such a waiver if the RFS requirement would severely harm the economy of a state, region, or the country. The EPA, in August, denied the request, stating that its analysis "found no compelling evidence that the RFS mandate is causing severe economic harm during the time period specified by Texas."[12] Opponents have argued that an EPA waiver, or reductions in the financial incentives proposed in the various bills, would not have much of an immediate impact on corn prices (see FAPRI and CARD studies, above). At any rate, other factors such as global grain shortfalls, strong U.S. exports, and higher oil prices played a much larger role in high feed costs, some have continued to assert.

Use Conservation Land

Some have argued that as much as one-third of the nearly 35 million acres (April 2008) in USDA's conservation reserve program (CRP) could be returned to agricultural production without environmental harm. The 2008 farm bill (P.L. 110-246) caps national enrollment in the CRP, which is intended to keep environmentally sensitive lands out of production for 10-15 years, at 32 million acres. What about incentives like easing producer penalties who want to end their CRP contracts early, discouraging re-enrollment or renewal of existing contracts, permitting temporary use of more CRP acres for grazing, cutting hay, and/or raising other feedstuffs? Opponents of opening conservation land assert that the loss of environmental benefits will far outweigh any gains to animal agriculture, in part because such lands have always been considered marginally productive in the first place. Also, much of this marginal CRP acreage previously was more suitable for (and planted in) wheat, not corn or soybeans.

USDA on May 27, 2008 announced the potential eligibility of more than 24 million acres enrolled in CRP for hay and forage in 2008 to help livestock producers hit by higher feed prices. USDA set a number of limitations on use of CRP acreage to minimize impacts on nesting birds and on the most environmentally sensitive lands. The National Wildlife Federation sued the Department, arguing that it did not prepare an environmental impact statement on the new policy. In July, a U.S. District Court issued a temporary restraining order (TRO) and then a permanent injunction suspending USDA's CRP opening, except for applicants that had been approved by or had applied prior to the TRO, or who had invested at least $4,500 toward haying or grazing equipment and preparation prior to the TRO. Where the application was approved prior to the TRO, haying and grazing must be completed by November 10. For subsequent approvals, haying and grazing must end by September 30 and October 15, respectively.[13]

Financial Assistance for Livestock Producers

Over the past decade alone, many of the nearly annual supplemental appropriations measures containing emergency agricultural aid have included money for animal agriculture — amounting to at least $3.2 billion in total, often to cover the cost of high feed prices or lost pasture use.[14] Separately, Section 32 of the Act of August 24, 1935 is a permanent appropriation available to USDA for a variety of purposes, and it has been tapped not only to pay for disaster losses, but also to address economic and market problems.[15] Thus, some argue, there is policy precedent for providing direct payments to livestock and poultry producers to help them cope with sharply rising feed costs. On the other hand, both the 2008 farm bill (P.L. 110- 246) and the FY2008 USDA appropriation (part of P.L. 110-161), contain language seeking to constrain Section 32 spending, or at least the Secretary's flexibility to use it.

Other Options

Some have urged U.S. officials to push harder for the elimination of various foreign barriers that obstruct international trade and distort market price signals to producers, including tariffs, unjustified technical and phytosanitary standards, export controls, and various domestic agricultural subsidies. Further, it is argued, investments should be increased, particularly in developing countries, in agricultural research and education to increase productivity, in transportation and marketing infrastructure, and in other types of alternative energy production here and globally. These efforts could all help to ease the upward pressure on feed grain and other commodity prices. Others counter that regardless of their merits, these types of solutions would offer little if any short-term relief to the financial problems currently facing livestock producers.

End Notes

[1] Unless noted, sources for this section include USDA, ERS, "Farm Income and Costs: 2008 Farm Sector Income Forecast," August 2008; USDA's *World Agricultural Supply and Demand Estimates*, September 12, 2008; testimony by USDA Chief Economist Joe Glauber before the Senate Committee on Energy and Natural Resources, June 12, 2008; and "Feed Grains and Livestock: Impacts on Meat Supplies and Prices," *Choices Magazine Online!*, 2nd quarter 2008.

[2] Cost of production data by John D. Lawrence and Shane Ellis, Iowa State University, accessed June 2008, at [http://www.econ.iastate.edu/faculty/lawrence/Lawrence_Website/estreturns.htm]. Lawrence has noted that these losses follow 10 years that were generally profitable for producers.

[3] LMIC website accessed on September 17, 2008, at [http://www.lmic.info/].

[4] "Something's Got to Give," *CattleFax Update*, June 20, 2008. Also see David P. Anderson et al., *The Effects of Ethanol on Texas Food and Feed*, Agricultural and Food Policy Center, Texas A&M University, April 10, 2008.

[5] National Milk Producers Federation (NMPF), July/August 2008 Dairy Market Report.

[6] Testimony of Joe L. Outlaw, Agricultural and Food Policy Center, Texas A&M University, before the Senate Committee on Energy and Natural Resources, June 12, 2008.

[7] Adapted in part from CRS Report RL34474, *High Agricultural Commodity Prices: What Are the Issues?*, where additional details, and (unless noted) sources may be found.

[8] USDA, ERS, *Feed Outlook*, September 16, 2008.

[9] Collins, Keith, *The Role of Biofuels and Other Factors in Increasing Farm and Food Prices*, June 19, 2008, study conducted in support of a review by Kraft Foods Global, Inc.

[10] Source: Renewable Fuels Association, June 19, 2008, at [http://www.ethanolrfa.org/].

[11] "Figures Reveal How Livestock Benefit From Ethanol," *The PigSite*, accessed August 18, 2008, at [http://www.thepigsite.com/].

[12] "EPA Keeps Biofuels Levels in Place after Considering Texas' Request," August 7, 2008 press release, which can be found, along with a fact sheet on the EPA findings and the August 13, 2008 *Federal Register* notice, at [http://www.epa.gov/otaq/renewablefuels/].

[13] See CRS Report RS21613, *Conservation Reserve Program: Status and Current Issues*, by Tadlock Cowan. The order does not affect a July 2008 USDA announcement to permit grazing only on CRP land in designated flood-stricken areas.

[14] The 2008 farm bill (P.L. 110-246) creates a new Agricultural Disaster Relief "Trust Fund" for crop years 2008-2011. Three of the five new programs under which payments could be made are relate to livestock. See CRS Report RL34207, *Crop Insurance and Disaster Assistance in the 2008 Farm Bill*, and CRS Report RL33958, *Animal Agriculture: 2008 Farm Bill Issues*.

[15] See CRS Report RL34081, *Farm and Food Support Under USDA's Section 32 Program*.

CHAPTER SOURCES

Chapter 1 - This is an edited, reformatted and augmented version of a Congressional Research Service publication, R41956, dated August 11, 2011.

Chapter 3 - This is an edited, reformatted and augmented version of a USDA publication, LDP-M-170-01, dated September, 2008.

Chapter 4 - This is an edited, reformatted and augmented version of a Congressional Research Service publication, RS22908, dated September 17, 2008.

INDEX

D

E

F

S

T

V

W

Y